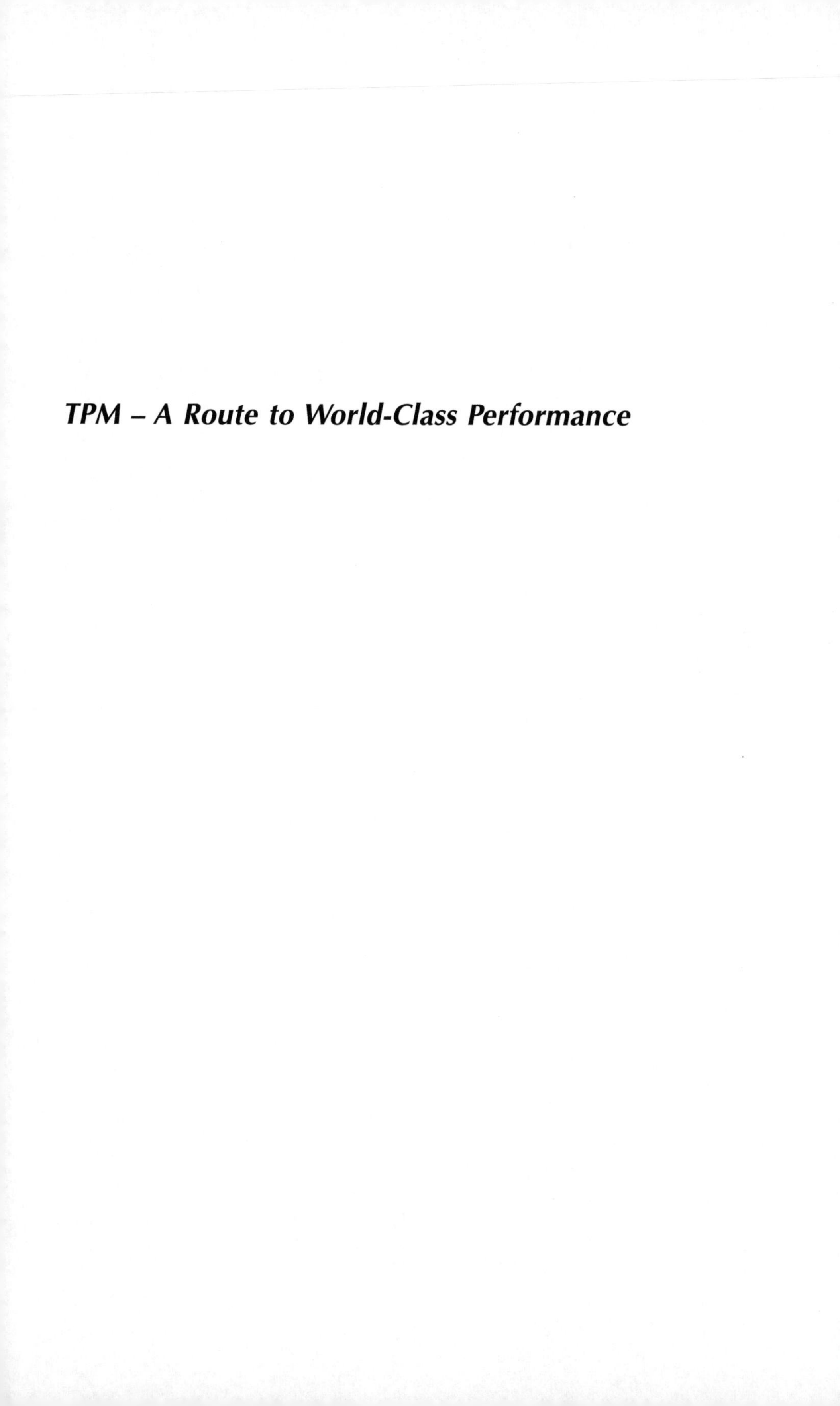

TPM – A Route to World-Class Performance

TPM – A Route to World-Class Performance

Peter Willmott and Dennis McCarthy

Oxford Auckland Boston Johannesburg Melbourne New Delhi

Butterworth-Heinemann
Linacre House, Jordan Hill, Oxford OX2 8DP
225 Wildwood Avenue, Woburn, MA 01801-2041
A division of Reed Educational and Professional Publishing Ltd

A member of the Reed Elsevier plc group

First published 2001

British Library Cataloguing in Publication Data
A catalogue record for this book is available from the British Library

Library of Congress Cataloguing in Publication Data
A catalogue record for this book is available from the Library of Congress

ISBN 0 7506 4447 8

Typeset at Replika Press Pvt Ltd, 100% EOU, Delhi 110 040, India
Printed and bound in Great Britain by
Biddles Ltd, Guildford and King's Lynn

Contents

Foreword

The global market means we all have to reach world-class levels of performance. Those of us that don't will be left behind. Big or small, private or public; there's no hiding place from the winds of change.

That's why the AEEU recognized the value of TPM some years ago. We saw then that TPM could enable manufacturing and services industries to become the best in the world.

Unlike TQM, which was conceptually sound, but patchy in outcome, TPM offers a new and invigorating approach. Involving everyone from shopfloor to boardroom, TPM is a team-based and freshly focused tool for success.

And it can really help Partnership work. In many ways, it's the missing link. TPM shows that cost reduction doesn't just mean redundancy. Rather, it offers the opportunity for managers and the workforce to share the gains from reducing waste and boosting productivity and performance.

Two years ago, I launched a new initiative called 'Partners in Success'. Built around the principles of TPM, it showed how unions and employers could raise their game and then share the rewards. And that message hasn't changed.

And it's a pleasure to recommend another WCS publication on TPM. The AEEU has worked closely with Peter Willmott, Dennis McCarthy and others for some time now. We've been spreading the TPM message – at the DTI, throughout industry and at conferences and seminars in the UK and across Europe.

That's why this book will help organizations meet the challenges of the new knowledge economy. It is a tool for change and a guide through new markets. It ought to be on every practical manager's bookshelf.

Sir Ken Jackson
General Secretary
AEEU

Preface

Customers expect manufacturers to provide excellent quality, reliable delivery and competitive pricing. This demands that the manufacturer's machines and processes are highly reliable. But what does the term 'highly reliable' really mean?

Certainly, with manufacturing, process and service industries becoming progressively dependent on the reliability of fewer but more sophisticated machines and processes, it means that poor equipment operating performance is no longer affordable or acceptable. The overall effectiveness of our machines, equipment and processes is paramount to provide consistency of product quality and supply at a realistic price.

Coping with modern manufacturing technology that is intrinsic in the materials, mechanisms and processes which we invent, design and use is one issue. Delivering the manufacturing company's vision and values as a lean, just-in-time producer to its customers, shareholders and employees is another.

Some world-class Japanese companies recognized over twenty-five years ago that the effective application of modern technology can only be achieved through people – starting with the operators and maintainers of that technology – and not through systems alone. Hence the emergence of total, productive maintenance as the enabling tool to maximize the effectiveness of our equipment by setting and maintaining the optimum relationship between people and their machines.

The problem with the words 'Total Productive Maintenance' – and hence the philosophy or technique of TPM – is that, to Western ears, they sound as though TPM is a maintenance function or a maintenance department initiative. But it is *not!* On the contrary, TPM is driven by manufacturing which picks up production and maintenance as equal partners: it is no longer appropriate to say 'I operate, you fix' and 'I add value, you cost money'. What TPM promotes is: 'We are *both* responsible for this machine, process or equipment and, between us, *we* will determine the best way to operate, maintain and support it'. Perhaps a better way of describing TPM, therefore, is to think of it as Total Productive Manufacturing, as it picks up operations and maintenance as equal partners under the umbrella of manufacturing.

The problem of definition has arisen because the word 'maintenance' has a much more comprehensive meaning in Japan than in the Western world. If you ask someone from a typical Western manufacturing company to define the word 'maintenance', at best he might say, 'Carry out planned servicing at fixed intervals'; at worst he might say, 'Fix it when it breaks down'. If you ask a Japanese person from a world-class manufacturing company, he will probably say, 'Maintenance means maintaining and *improving* the integrity of our production and quality systems through the machines, processes, equipment and people who add value to our products and services, that is, the operators *and* maintainers of our equipment'. Whilst this may be a longer definition, it is also a more comprehensive and relevant description. Hence it is now more appropriate to think of TPM as Total Productive Manufacturing.

Over the last few years, certainly since the advent of the 1990s, a growing number of Western companies have, with varying degrees of success, adopted the Japanese TPM philosophy. The companies who have been successful in using TPM in their operations have recognized and applied some key success factors, including:

- You must enrol and secure the commitment of senior managers from the start.
- TPM is led by manufacturing.
- TPM is a practical application of total quality and teamwork.
- TPM is an empowerment process to give shared responsibility and ownership.
- The TPM philosophy is like a heart transplant: if you don't match it to the patient, you will get rejection. You must, therefore, treat each company or recipient as unique and *adapt* the principles of TPM to suit the local plant-specific issues without corrupting the well-founded and proven principles of TPM.

Total Productive Maintenance, an original Japanese management protocol developed to alleviate production losses caused by machine breakdowns, has moved on. Through TPM, more companies now accept the concept of zero breakdowns as achievable. From the foundation of striving for zero breakdowns, world-class plants are able to run for complete shifts without the need for intervention. TPM is still pushing back the boundaries of what was thought possible. This does not mean that people are no longer needed. On the contrary, it is the ingenuity of operators, maintainers, engineers and management, working as full members of the company team, which makes such progress possible, often working as a positive 'partnership for change'.

Based on our experience of working with world-class companies, this book provides a practical guide to delivering TPM benefits within cultures where professional cynics have had years to practise their craft. Based on the proven principles of TPM, the book emphasizes the need to build on existing good practices and to win commitment by delivering results. It is based on the author's first-hand experience of seeing TPM in Japan and then adapting those principles to suit the strategic needs of companies across four continents. It builds on Peter's earlier book *TPM the Western Way*, updating the scope of applications and tools. It includes more detail on the 'life after pilot' as well as the application of TPM to equipment design, administration and non-manufacturing areas. The TPM route map is updated to include the journey to zero breakdowns and beyond. It also provides a systematic structure to evolve from the classic Total Productive Maintenance towards Total Productive Manufacturing and, hence, deliver a Totally Productive Operation capable of world-leading performance.

Peter Willmott
Dennis McCarthy

Acknowledgements

Since the first edition of this book some five years ago, we have had the privilege of working with some really good people who share our passion and belief in the potential power of the TPM philosophy.

This shared passion came from touching TPM in practice rather than theory. So here, we would like to express our sincere thanks to some of those good people.

Our many clients continue to be our main source of inspiration through the priority, pace and resources they apply to the TPM process and their operations to gain and sustain competitive advantage. We are particularly indebted to:

Derek Cochrane and Derek Taylor of 3M
Chris Rose and Clive Marsden of Adams
Les Thomson and Mike Milne of BP Amoco
Ian Barraclough of Elkes Biscuits
Gordon Hill and Mike Williamson of Henkel
Danny McGuire and Ken O'Sullivan of RHP Bearings
Grant Budge and many others at RJB Mining

The UK's Department of Industry, specifically through Peter Dalloway, Robin Crosher, John Gillies and Richard Arnott, has actively supported practical research, such as the 'TPM Experience' project aimed at identifying and disseminating 'TPM Best Practice'. This provides an excellent example of government working in partnership with industry, and long may it continue!

In spreading the TPM message via conferences, workshops and study tours, the roles played by David Willson of Conference Communication and John Moulton and Tom Brock of Network Events, plus John Dwyer and Paddy Baker of Findlay Publications, are all acknowledged with gratitude. Also Doug Osman, formerly a Team Leader at Hoechst Trespaphan, has been an avid supporter of the TPM philosophy at many of our workshops over the years.

A special word of thanks is due to our colleagues in WCS International who have provided much of the inspiration, perspiration and material for this book, and to Heather Scott-Duncan who undertook the onerous task of word-processing the text.

Particular thanks also to Sir Ken Jackson, General Secretary of the AEEU, for his kind contribution in writing the foreword to this book, and to Lynn Williams, National Officer, AEEU, who has been an effective and influential missionary for TPM over the last several years.

Last, but not least, our gratitude to our wives for having the patience to support us and for creating an environment which has given us the time to write this book.

Glossary of TPM terms

Asset Care Programme: A systematic approach to keeping equipment in 'as new' condition. This consists of carrying out routine activities such as: cleaning and inspection (carried out by the operator and sometimes called Operator Asset Care or Autonomous Maintenance), checking and monitoring (sometimes called Condition Based Monitoring), preventative maintenance and servicing (sometimes called Maintainer Asset Care).

Availability: The actual run time of a machine as a percentage of its planned run time.

Best of the Best: An OEE figure calculated by multiplying the best weekly availability, the best weekly performance and the best weekly quality rates for a machine over a period, e.g. of typically one month.

CAN DO/5S: Five common sense principles of workplace organization (arrangement, neatness, cleaning, order and discipline). CAN DO is the western equivalent of the Japanese 5Ss.

Condition Appraisal: The assessment of the condition of a machine's components as a first step to undertaking the refurbishment plan and improving the OEE. This must involve carrying out a deep clean as part of the assessment.

Condition Cycle: The second stage of the TPM Improvement Plan, which includes criticality assessment, condition appraisal, refurbishment plan and the asset care programme.

Core Team: These are the mixed shift based teams comprising operators and maintainers and a Team leader. These teams work through the 9 Step TPM Improvement Plan on their Pilot Projects typically over a 12 to 16 week period.

Criticality Assessment: An evaluation of each of the machine's components against set criteria and their likely impact on production, safety, environment and cost.

Five Whys: Asking 'Why?' five times to try to get to the root cause of the problem.

Four Milestones: This is the progression the organization goes through over a period of approximately 4–6 years as they embark on the TPM Process. These have been recognized as discrete phases that organizations go through as they transform themselves. The 4 Milestones are:

1. Introduction
2. Refine Best Practice and Standardize
3. Build Capability
4. Zero Losses

At each Milestone there will be audits to establish your capability in transforming the business and further planning to take into account the future changes required to meet customer and market needs, as well as the organization's needs. For each milestone your management team will have defined goals and targets as Pillar Champions, that should be realized having reached each of the milestone stages.

Improvement Zone (IZ): This is a geographical area where the First Line Managers and their teams apply the basic techniques of TPM and CANDO/ 5S. This area is a manageable but representative portion of the process or plant which when improved, will provide an important contribution to the business.

Key Contact: These are support personnel, usually from the functional departments like Finance, Design, Engineering, Laboratory, or individuals with specialist knowledge. They will gradually get involved either with an improvement project for themselves or using their specialist knowledge to support an improvement team. Their aim is to support organizational learning and problem resolution using the tools of TPM.

Maintainability: This refers to how easy it is to gain access to the equipment and the particular skills needed to diagnose a problem.

Measurement Cycle: The first stage of a TPM Improvement Plan, consisting of collecting equipment history, calculating the OEE and assessing the Six Losses.

Minor Stoppage: When a machine is stopped for a relatively short period (e.g. to clear a blockage) and then re-started without the need for any repair. A minor stop therefore causes an Operator to have to interfere with the process, but does not require the attendance of a Maintenance Technician.

Nine Step TPM Improvement Plan: This is a set of steps the Core Teams progressively go through when analysing the Pilot Plant/Area. It enables them to understand the equipment, measure the problems, analyse then fix the condition of the equipment and lastly pass on specific technical or support issues still to be resolved. By doing so the teams will improve the equipment, but more importantly they will discover the real reasons why the equipment is in the condition we see it and why it's not performing in the way we would want. Some of these issues can be fixed quickly and some are more long term. Only the critical plant and equipment will be subject to the 9 Step Improvement Plan.

OEE: A measure used in TPM to calculate the percentage of actual effectiveness of the equipment. Taking into consideration the availability of the equipment, the performance rate when running and the quality rate of the manufactured product measured over a period of time (days, weeks or months). The difference between the current OEE and its maximum potential is the current cost of non-conformity. Sometimes called the 'hidden factory'.

Operational Improvements: Improvement activities which result in increasing the equipment's reliability once implemented by the TPM core team. The objective being to make it easy to do things right and difficult to do things wrong.

Pillar Champions: Initially there are five very important capabilities that everyone needs to embrace if TPM is to flourish. These are:

1. Continuous Improvement in OEE (OEE)
2. Maintenance Asset Care (MAC)
3. Operator Asset Care (OAC)
4. Skills Development (SD)
5. Early Equipment Management (EEM)

Because the five principles (sometimes called the Pillars of TPM) are so important we assign their development to each member of the management team. Each Pillar Champion as they are referred to creates the environment at the Plant, by changing the way they manage, to enable everyone to contribute to these principles and the TPM process. They therefore develop the policy for the particular pillar and then ensure its consistent deployment.

Pilot Projects: These initial pilots are learning experiences for the core teams to work through the nine Step TPM Improvement Plan. They are small but representative 'chunks' of plant that enable us to flush out the management processes and habits that need to change if we want TPM to flourish across a plant or site.

P–M Analysis: A problem solving tool used in TPM in conjunction with the five whys. The 4 Ps and the 4 Ms stand for:
4 Ps – phenomena which are physical in nature which cause problems that can be prevented.
4 Ms – are caused by machines, manpower, methods and materials.

Performance Rate: The actual performance rate of a machine or process, expressed as a percentage of planned performance rate.

Problem Prevention Cycle: The third and final stage in the Improvement Plan when the TPM Core Team concentrates on preventing problems from occurring in the future.

Roll-Out: This is where we start implementing the TPM techniques across the whole site. This is so that we can begin to get everyone involved and contributing to the TPM process. This also has a number of stages (called Phases). These are staggered so that we implement TPM at a sustainable rate.

Quality Loss: Lost production due to the manufactured product not being produced right first time and which will therefore need to be either re-worked or scrapped.

Quality Rate: The first time ok product, expressed as a percentage of the total manufactured.

Reduced Speed Loss: Production lost due to running equipment at a speed lower than its intended or designed speed.

Refurbishment Plan: Identifying all the activities that need to be undertaken in order to restore the equipment to 'as new' condition. This includes an estimate of the cost, manpower resources, agreed priorities, timing and responsibilities.

Scoping Study: The Scoping Study provides information to support the development of the TPM implementation programme. This will include a cost/benefit appraisal. It also identifies any potential roadblocks and provides an indication of the workforce's perception and feelings.

Set Up and Adjustment Losses: Production time lost because a machine is being set up or adjusted at the start of a run.

Six Losses: These are the categories of losses the TPM teams use to identify and measure plant problems so that they can prioritise them and progressively reduce or eliminate them. These are the things that affect your Overall Equipment Effectiveness (OEE) score. The six Losses are:

1. Breakdowns
2. Excessive Set Up and Changeover
   These two affect whether the machine is available to produce or not. This is why we use this as the AVAILABILITY percentage within the OEE calculation
3. Idling and Minor Stops
4. Running at Reduced Speed
   These two affect the PERFORMANCE of your machine when running. This is the percentage rate within the OEE calculation.
5. Reduced Yield (Scrap & Rework)
6. Start Up Loss

These two affect the Quality of the product produced on the machine. This is the QUALITY percentage within the OEE calculation.

Start Up Loss: Lost production due to defects which occur at the start of a run.

Support Improvements: Improvements to equipment efficiency that can only be achieved through changes in other parts of the organization.

Support Team: This team includes representatives from each support function such as finance, design, engineering, production control, quality control, supervision, and a Union representative. Usually referred to as Key Contacts.

Technical Improvements: Improvements to equipment efficiency that require technical analysis of problems before they can be implemented.

TPM: TPM is the abbreviation of Total Productive Maintenance. It is a comprehensive strategy that supports the purpose of equipment improvement to maximize its efficiency and product quality. Many TPM practitioners prefer to call it Total Productive Manufacturing to highlight the need for an equal partnership between production and maintenance.

1

Putting TPM into perspective from Total Productive Maintenance to Total Productive Manufacturing

If we look at the value stream (Figure 1.1), we see that it is customers who actually drive our business. In the manufacturing sense, we therefore need to provide the necessary production responses to satisfy and exceed those expectations by adding value, quality and performance in all that we do.

The most effective way of adding value is to have a continuous determination to eliminate waste across the supply chain and thus maximize the value stream: easy to state, difficult to deliver.

So where do the principles, processes and reality of TPM come into play to achieve the goal of a 'Totally Productive Operation'?

1.1 TPM applied company-wide

The answer is to view TPM not simply as Total Productive Maintenance in

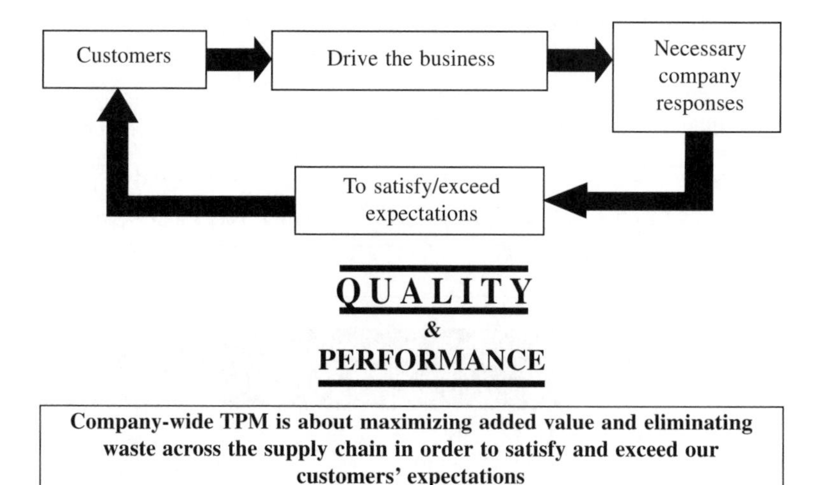

Figure 1.1 The value stream

the sense of Overall Equipment Effectiveness (OEE), autonomous maintenance, 5 Ss, clean machines and so on, but rather as the proven roots and origins for applying company-wide TPM (see Figure 1.2).

The original fifth pillar of TPM, Early Equipment Management or TPM for Design, links well with the broader view that TPM stands for Total Productive Manufacturing. As such, it is not a Maintenance Department-driven initiative, but actually brings production and maintenance together as equal partners under the umbrella of manufacturing.

Similarly, 'TPM in the Office' is better served by broadening the application of these sound and proven principles into 'TPM in Administration', embracing all support functions such as sales, marketing, commercial, planning, finance, personnel, logistics, stores and information technology (IT).

Company-wide TPM recognizes that:

- if equipment OEE improves but the overall door-to-door time remains the same, the waste is not removed;
- if equipment capability improves but quality standards remain the same, a potential area of competitive advantage is lost;
- if knowledge gained about the process does not produce higher rates of return on investment, the organization is not making the best use of its capabilities;
- if capability is increased but this is not met by generation of new business, an opportunity to reduce unit costs is lost.

1.2 Presenting the business case: what is Overall Equipment Effectiveness?

The true costs of production are often hidden. TPM addresses an 'iceberg' (Figure 1.3) of supply chain losses. Secondly, total life cycle costs can be more

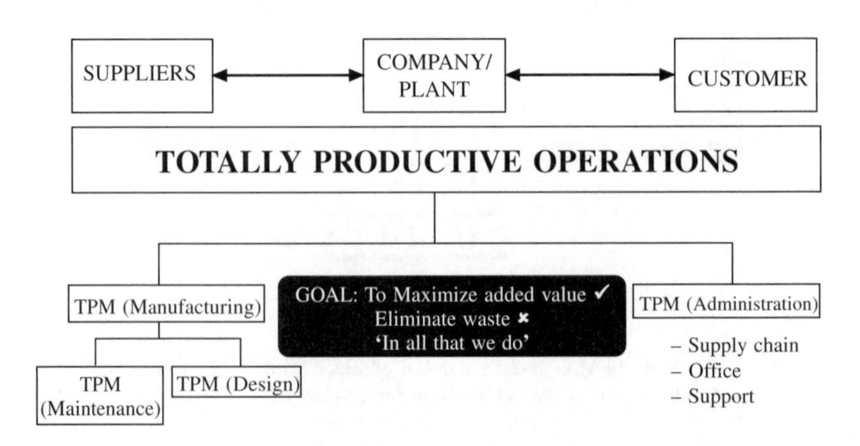

Figure 1.2 The value stream and TPM

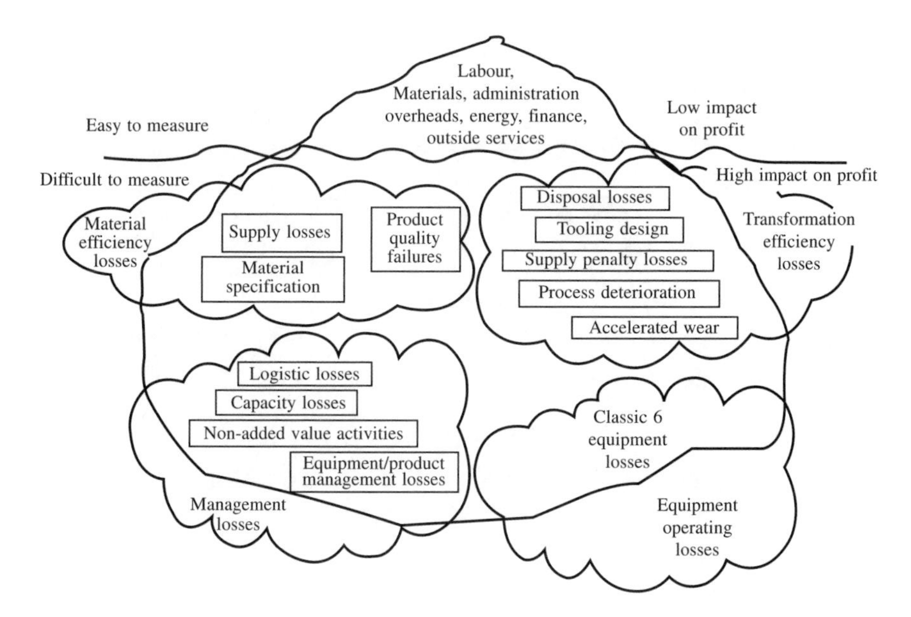

Figure 1.3 Supply chain hidden losses

than twice as high as the initial purchase price. Through TPM the useful life of equipment is extended; you can therefore get more from your investment. Thirdly, if capacity can be increased to consistently achieve its design potential, then the fixed cost per unit will be significantly, and quite often many times, reduced.

1.3 Attacking the hidden losses

Many companies attack the direct, visible costs without considering the lost opportunity hidden costs.

To do either in isolation is both narrow and ineffective. What TPM does is to attack the hidden losses *and* ensure value for money from the direct manufacturing effort. The combined strategy will result in a dramatic benefit. This approach is sometimes called 'Cost deployment using TPM'. It could more appropriately be called Loss deployment, as it focuses on both cost and opportunities for added value.

Company-wide TPM starts by attacking the six classic shopfloor losses affecting equipment effectiveness (see page 5). This is the main focus of shopfloor teams. Figure 1.4 shows the relationship between these losses to be addressed at each management level and the added value in terms of increased competitiveness.

The volume of throughput is a key determinant of unit cost. It is easy for management under pressure to concentrate on satisfying current demand rather than growing future business.

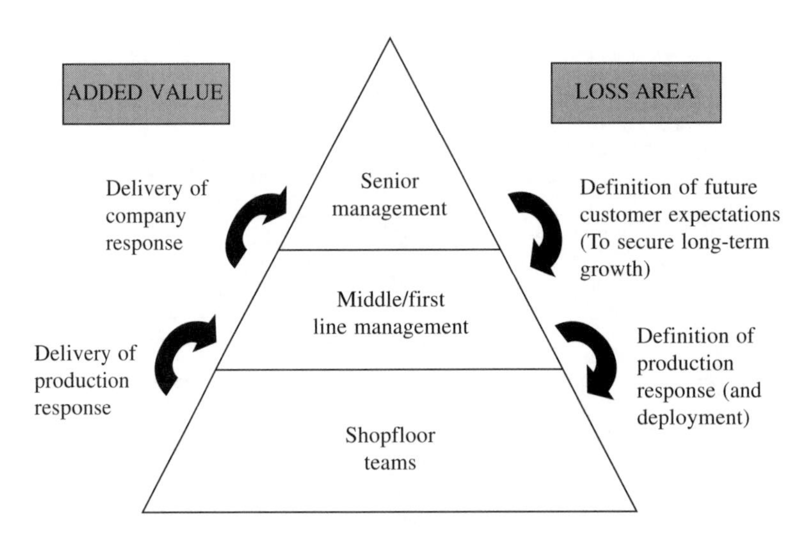

Figure 1.4 The value stream

Establishing this company-wide perspective places equal emphasis on strategic direction *and* delivery. This helps managers at all levels to present a consistent set of leadership values and behaviours – an effective countermeasure to a shopfloor battleground littered with incomplete initiatives. Systematically applied TPM is also strong enough to sustain direction as managers progress through their career development path (the most common root cause of initiative fatigue).

The key is to measure and monitor all the major hidden losses, then direct the company resources to reducing those that will increase the organization's profitability.

1.4 OEE and business success

The classic measure of Overall Equipment Effectiveness is the product – in percentage terms – of the *Availability* of a piece of equipment or process × its *Performance Rate* when running × the *Quality Rate* it produces.

The OEE measure is not just limited to monitoring the effectiveness of a piece of machinery, however. It can, and should be, applied to the business as a whole, assessing the productivity of the complete value chain from supplier to customer.

A very powerful business measure, the OEE is a key performance indicator (see Figure 1.5) that can be applied at the three key levels.

- *Shopfloor* – the machine or process: – floor-to-floor OEE

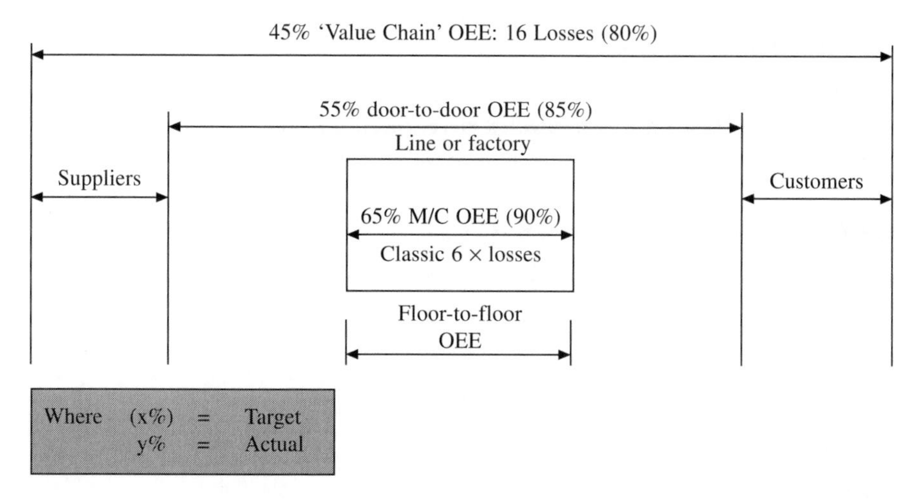

Figure 1.5 OEE: Key performance indicator

- *First-line management* – the line or factory: door-to-door OEE
- *Senior management* – the business, i.e. supplier to customer: value chain OEE.

As the figure illustrates, there is little merit in driving up the machine OEE from 65 per cent to 90 per cent by attacking the classic six losses, if the door-to-door, line or factory OEE stays at 55 per cent. Similarly, you will not satisfy your customers if the value chain OEE remains at 45 per cent.

As stated earlier, company-wide TPM is concerned about attacking all forms of waste. In the illustration, measurement of the machine OEE will allow the operator/maintainer core TPM team to focus their efforts on prioritizing and then attacking the classic six losses of:

- Breakdowns
- Set-ups and changeovers
- Running at reduced speeds
- Minor stops and idling
- Quality defects, scrap, yield, rework
- Start-up losses

The first two losses affect *availability* the second two affect the *performance* rate when running, and the final two affect the *quality* rate of the resultant OEE figure.

This measurement must also highlight the door-to-door losses outside their immediate control, so that first-line management can prioritize the flows to and from the machine. This form of door-to-door measure will typically highlight the following.

To the machine or process

- Ineffective raw material and tool marshalling
- Lack of forklift and/or forklift driver availability
- Inadequate access to the machine or process

From the machine or process

- Inadequate take-off/take-away facilities (i.e. track reliability)
- Upstream/downstream bottlenecks
- Poor shift handover arrangements

Finally, the value chain OEE measure is a senior management key performance indicator, typically aimed at highlighting the following.

Suppliers

- Poor procurement procedures
- Poor quality and/or lack of consistency of incoming materials/ components

Customers

- Lack of responsiveness to customer call-off changes

In our experience, part of the essential planning and scoping stage prior to TPM implementation is a detailed assessment of the three OEE levels outlined above. Do not be surprised if well over half of your lost opportunity costs or costs of non-conformity lie outside the machine or process OEE.

In reality, overall equipment effectiveness measures how well a company's production process or individual piece of equipment performs against its potential. Through the best of best calculation, it also indicates a realistic and achievable target for improvement.

Not only that, but due to the linkages with the hidden losses, it identifies the technique which can best address the type of problem. (Each loss has a different TPM approach to resolving it.)

The outcome of a *low* OEE is a reactive management style. Here the root cause of many of the unplanned events throughout the organization can be traced back to the production process. As OEE is raised through TPM, opportunities are presented to drive out waste and improve customer service.

The route to TPM encourages teamwork and cross-functional learning. As such, TPM provides a mechanism to deliver change when directed correctly, which can have a powerful impact on company-wide perceptions and attitudes.

An improving OEE indicates:

- how successful the organization is at achieving what it sets out to do;
- success in establishing a continuous improvement habit;
- buy-in to the company vision and values;

- development of capability to achieve and then exceed current accepted levels of world-class performance.

Some myths and realities of the OEE measurement process can be described as shown in Table 1.1.

Table 1.1 Myths and realities of OEE

Myth	*Reality*
OEE is a management tool to use as a benchmark	This misses the benefit of OEE as a shopfloor problem-solving tool
OEE should be calculated automatically by computer	The computation approach is far less important than the interpretation. While calculating manually, you can be asking why?
OEE on non-bottleneck equipment is unimportant	OEE provides a route to guide problem solving. The main requirement is for an objective measure of hidden losses even on equipment elsewhere in the chain
OEE is not useful because it does not consider planned utilization losses	OEE is one measure, but not the only one used by TPM. Others include productivity, cost, quality, delivery, safety, morale and environment
We don't need any more output, so why raise OEE	Management's job is to maximize the value generated from the company's assets. This includes business development. Accepting a low OEE defies commercial common sense

1.5 Modern role of asset care and TPM

Although TPM is better explained as Total Productive Manufacturing, the way in which maintenance is perceived is a key indicator of a world-class perspective.

How does 'asset care' impact on the business drivers and hence the OEE, productivity, cost, quality, delivery, safety, morale and the environment?

In the world-class manufacturing companies there is one common denominator: a firm conviction that their major *assets* are their machines, equipment and processes, together with the people who operate and maintain them. The managers of these companies also recognize a simple fact: it is the same people and equipment that are the true wealth creators of the enterprise. They are the ones that add the value. TPM is about asset care, which has a much more embracing meaning than the word 'maintenance'.

The traditional approach to industrial maintenance has been based on a functional department with skilled fitters, electricians, instrument engineers and specialists headed by a maintenance superintendent or works engineer. The department was supported by its own workshop and stores containing spares known from experience to be required to keep the plant running. The maintenance team would take great pride in its ability to 'fix' a breakdown or failure in minimum time, working overnight or at weekends and achieving the seemingly impossible. Specialized spares and replacements would be held in stock or squirrelled away in anticipation of breakdowns.

In the period after the Second World War this concept of breakdown maintenance prevailed. It was not until the 1960s that fixed interval overhaul became popular; this entailed maintenance intervention every three months or after producing 50 000 units or running 500 hours or 20 000 miles. The limitation of the regular interval approach is that it assumes that every machine element will perform in a stable, consistent manner. However, in practical situations this does not necessarily apply. There is also the well-known syndrome of trouble after overhaul: a machine which is performing satisfactorily may be disturbed by maintenance work, and some minor variation or defect in reassembly can lead to subsequent problems.

It is interesting to consider some statistics of actual maintenance performance in the early 1990s. Much of the material quoted below has been derived from a survey carried out by the journal *Works Management* based on a sample of 407 companies in 1991.

Expenditure on maintenance in the European Union (EU) countries has been estimated at approaching 5 per cent of total turnover, with a total annual spend of between £85 billion and £110 billion. This spend is equivalent to the total industrial output of Holland, or between 10 per cent and 12 per cent of EU industries' added value. Some 2 000 000 people in 350 000 companies are engaged in maintenance work (Table 1.2).

When we look specifically at the UK, we find the annual spend in 1991 was £14 billion, twice the UK trade deficit at that time or 5 per cent of annual turnover. It also equates with three times the annual value of new plant investment in 1991 or 18 per cent of the book value of existing plant (Table 1.3).

Table 1.2 Maintenance expenditure as a percentage of turnover in EC countries

UK	5.0%
France	4.0%
Italy	5.1%
Spain	3.6%
Ireland	5.1%
Holland	5.0%

Table 1.3 UK maintenance spending

- £14 billion annual spend
- Twice UK trade deficit
- 5% of sales turnover
- Three times value of new plant investment
- 18% of book value

Figure 1.6 gives an indication of the range of maintenance costs in various UK industries expressed as a percentage of the total manufacturing costs. The lowest band is around 5 per cent for the electrical, electronic and instrument industries, and the highest averages 14 per cent for the transportation industry.

At the time of the *Works Management* survey (1991), the technique most widely employed (40 per cent of companies surveyed) was running inspection. This was followed by oil analysis (27 per cent), on-line diagnosis (25 per cent) and vibration analysis (20 per cent). Fixed cycle maintenance and reliability-centred maintenance came next, but were in their infancy, indicating the enormous scope for the application of TPM to UK industry.

Finally, we look at the scope for moving from unsatisfactory to satisfactory maintenance. The pie charts in Figure 1.7 show the potential in moving away

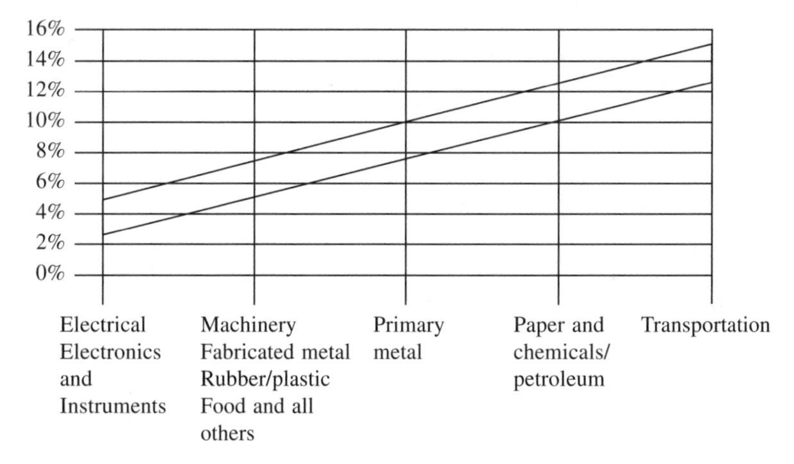

Figure 1.6 Maintenance spend as percentage total manufacturing cost

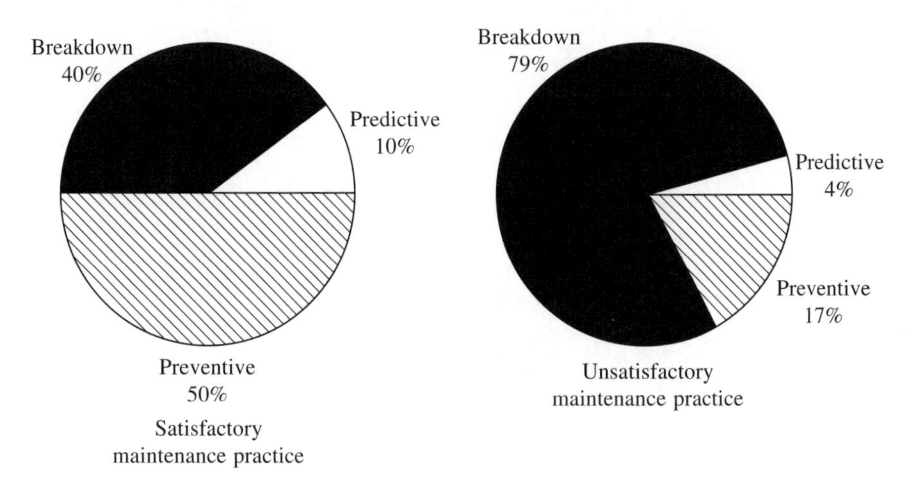

Figure 1.7 UK type of maintenance. Source: Works Management, July 1991

from breakdown and towards predictive and preventive approaches. The histogram chart in Figure 1.8 serves as an indication of objectives for improvements in performance from the present unsatisfactory levels to benchmark levels.

Japanese methods have been at the heart of the transformation in manufacturing efficiency which has taken place over the last twenty to thirty years and which is still going on. There are common threads running through all of these methods:

- Developing human resources
- Cleanliness, order and discipline in the workplace
- Striving for continuous improvement
- Putting the customer first
- Getting it right first time, every time

Central to all these approaches to manufacturing efficiency is the concept of TPM. Asset care has to become an integral part of the total organization so that everyone is aware of, and involved in, the maintenance function. The end result is that breakdowns become a positive embarrassment and are not allowed to occur. The assets of the production process are operated at optimum efficiency because the signs of deterioration and impending failure are noticed and acted upon.

Asset care covers three interrelated issues combining autonomous maintenance and planned preventive maintenance.

- *Cleaning and inspection*: Daily activities to prevent accelerated wear
- *Checks and monitoring:* Early problem detection and diagnosis
- *Preventive maintenance and service:* Injection of relevant technical expertise to prevent failure and restore condition

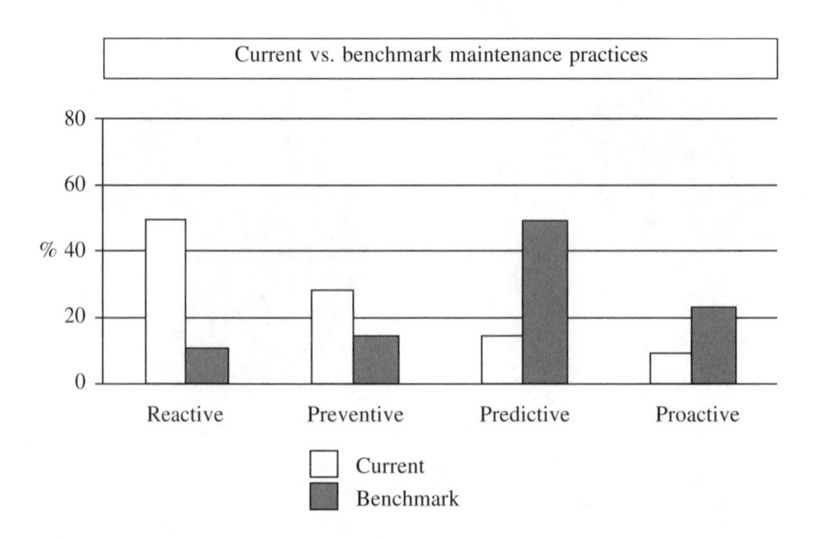

Figure 1.8 Asset care balance

These are described in more detail in Chapter 3. It should be recognized that asset care is something that evolves with experience. Once established, it is refined to reflect the improved equipment condition.

First sporadic losses

The route to high levels of reliability is reasonably predictable. Asset care is directed first towards sporadic losses. These are sudden failures such as breakdowns. They can be almost totally removed by improving equipment condition, reducing human error through training, fool-proofing and establishing how to detect potential failures before they occur. Experience shows that effective asset care can detect 80 per cent of potential component failures and stabilize the life span of the remaining 20 per cent. This is why zero breakdowns are becoming an accepted reality in most world-class operations.

Then optimization

Once sporadic losses are under control, the target becomes chronic losses. These require improved problem-finding skills. As such, asset care is refined to look for minor quality defects which direct first the definition and then the implementation of optimum conditions.

The seven steps of autonomous maintenance provide the route map to this evolution but, to be effective, must be supported by similar restructuring of planned maintenance activities.

These two activities are the core of the improvement zone implementation process described in Chapter 8. Teams translate management standards into local policy/best practice covering the following:

- Basic systems of problem detection, including initial cleaning and information, to understand the root causes and develop countermeasures;
- Basic lessons of maintaining equipment condition and increased understanding of equipment functions to correct design weaknesses and systemize asset care;
- Standardize and practise to achieve zero breakdowns and then optimum conditions.

The senior management role in asset care

The pace of progress through these stages is directly related to the priority which management assigns to it. To simplify this effect, TPM identifies clear management roles. These roles, known as *pillar champions*, provide leadership in terms of:

- setting priorities (where to start, what next?)
- setting expectations (work standards to be applied consistently)
- giving recognition (reinforce values)

The pillar champion roles and their relationship with the rest of the TPM infrastructure is set out in Figure 1.9.

Experience shows that for every breakdown there are many contributory factors. These factors are common to more than one breakdown. As such, analysis of breakdowns, even where the specific event is a rare occurrence, will highlight countermeasures which will prevent problems. The pillar champions have the top-down perspective to maximize this.

Asset care is, therefore, the engine-room of continuous improvement, providing:

- evidence of change
- increased understanding and ownership of equipment
- problem-finding resources
- a mechanism for locking in gains and problem prevention
- a means of reinforcing expectations and raising standards

A commitment to reliable asset care is the first level of management behaviour required to support continuous improvement.

1.6 5S/CAN DO philosophy

One of the first and crucial steps towards asset care comes from the application

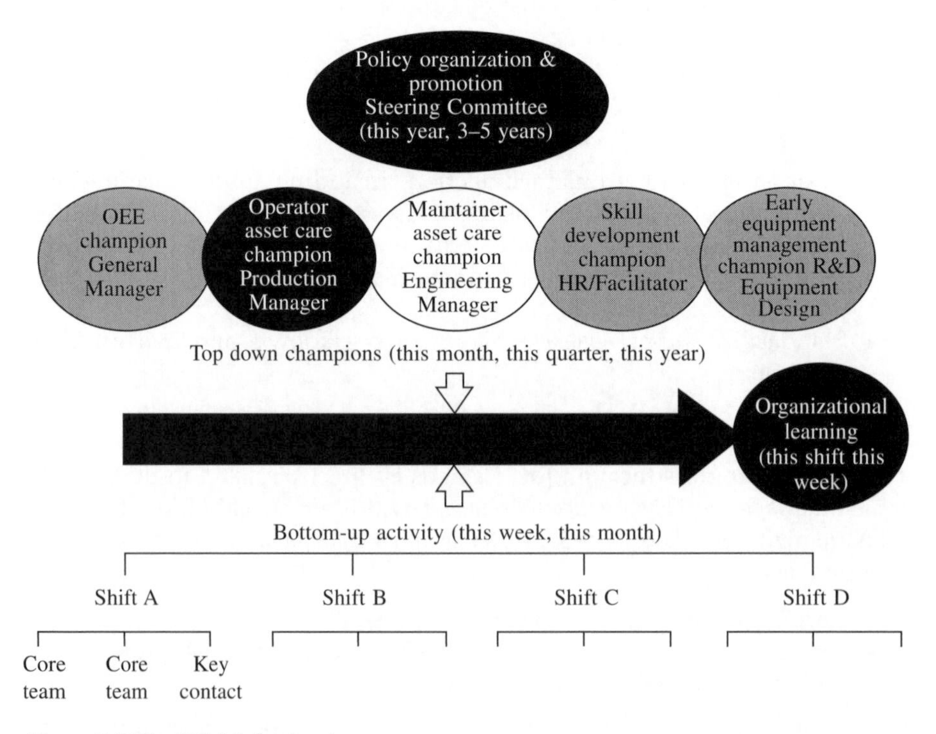

Figure 1.9 The TPM Infrastructure

of the 5 Ss, which are central to all the Japanese methods evolved since the end of the Second World War:

- *seiri* organization
- *seiton* orderliness
- *seiso* cleaning (the act of)
- *seiketsu* cleanliness (the state of)
- *shitsuke* discipline (the practice of)

In English-speaking countries an alternative way of expressing the 5 Ss is the more easily remembered CAN DO of:

- cleanliness
- arrangement
- neatness
- discipline
- order

The philosophy is exactly the same, however:

1. Get rid of everything and anything unnecessary.
2. Put what you do want in its right place so that it is to hand.
3. Keep it clean and tidy at all times, recognizing that cleanliness is neatness (a clear mind/attitude), is spotting deterioration (through inspection), is putting things right *before* they become catastrophes, is pride in the workplace, giving self-esteem.
4. Pass on that discipline and order to your colleagues so that we *all* strive for a dust-free and dirt-free plant.

The CAN DO approach, therefore, is to look at the production facility and clean the workshop and its plant and machinery as it has never been cleaned before, whilst at the same time casting a ruthlessly critical eye at everything in the workplace. Nothing must be allowed to remain anywhere on the shopfloor unless it is directly relevant to the current production process. Good housekeeping thereafter becomes everyone's responsibility and a way of life.

The cleaning process involves the operators of machines and plant. As they clean, they will get to know their machines better; they will gradually develop their own ability to see or detect weaknesses and deterioration such as oil leaks, vibration, loose fastenings and unusual noise. As time goes on, they will be able to perform essential, front-line asset care and some minor maintenance tasks within the limits of their own skills. The process will take place in complete co-operation with maintenance people, who will be freed to apply their technical skills where needed.

With the attitude to cleanliness and good housekeeping understood, we can move on to explain the main principles on which TPM is founded. In Chapter 4 we explain the toolbox of techniques used to implement these principles and how to develop buy-in by developing understanding through practical application of the WCS nine-step TPM improvement plan.

1.7 Implementing TPM principles

The successful implementation of the five CAN DO steps provides a powerful organizational learning tool. This is because CAN DO influences two important areas of corporate memory:

- process layout
- best practice routines

It provides a positive development path for manager/shopfloor relationships, helping to highlight the barriers to change and conform to world-class values.

The TPM implementation process is built around the CAN DO steps (as are the seven steps of autonomous maintenance). This treats information, collation, equipment, understanding and maintenance as things that are necessary, compared to sources of contamination, human error and hidden losses as unnecessary items. Having decided what is necessary, work processes can then be formalized/refined.

One of the outcomes of implementing best practice in this way is that many tasks can be simplified such that they can be carried out by the most appropriate person. This releases specialist maintenance or production personnel to concentrate on optimization of plant and equipment, providing the gateway to 'better than' new performance. The stepwise implementation philosophy of the TPM principles is set out below.

Continuous improvement in OEE

The initial process of cleaning and establishing order leads to discovering abnormalities, and progresses through four steps:

1. Discover equipment abnormalities.
2. Treat abnormalities and extend focus to supply chain losses.
3. Set optimal equipment conditions to deliver future customer expectations.
4. Maintain optimal equipment conditions during delegation of routine management activities.

The objective of this process is to move progressively towards a situation where all production plant is always available when needed and operating as closely as possible to 100 per cent effectiveness. Achieving this goal will certainly not come easily and may take years. The basic concept is one of continuous improvement: 'What is good enough today will not be good enough tomorrow'.

Operator asset care (autonomous maintenance)

As operators become more closely involved in getting the very best from their machines, they move through seven steps towards autonomous or self-directed maintenance:

1. Initial cleaning
2. Carrying out countermeasures at the source of problems
3. Developing and implementing cleaning and lubrication standards

4 General inspection routines
5 Autonomous inspection
6 Organization and tidiness
7 Full autonomous maintenance

As these seven steps are taken, over an agreed and achievable timetable, operators will develop straightforward common-sense skills which enable them to play a full part in ensuring optimum availability of machines. At no stage should they attempt work beyond the limits of their skills: their maintenance colleagues are there for that purpose.

Maintainer asset care

In parallel with the operator asset care steps, maintenance best practice development supports the stepwise implementation.

1. Refurbish critical equipment and establish back-up strategies for software/systems.
2. Contain accelerated deterioration and develop countermeasures to common problems. Establish correct parameter settings.
3. Set condition monitoring and routine servicing standards to improve response times and reduce sporadic failures.
4. Use event analysis to fine-tune asset care delivery towards zero breakdowns.
5. Routine restoration of normal wear to stabilize component life. Hand over routine maintenance to operators.
6. Use senses to detect internal deterioration.
7. To extend component life and improve equipment life prediction.

Quality maintenance

The role of maintenance evolves from planned maintenance to lead the quality of maintenance in order to:

1. Eliminate accelerated deterioration.
2. Eliminate design weaknesses.
3. Eliminate minor quality defects as a route to delivering optimum conditions.
4. Systemize/fool-proof to maintain optimum conditions with reduced intervention.

Continuous skill development

The above will only become a reality provided we develop people's competences to:

- establish the purpose of training as a key lever for sharing ideas, values and behaviours;
- establish training objectives linked to business goal delivery;
- agree methods of delivery which make it easy to deploy ideas cross-shift;

- set up a training framework and modules to systematically build capability;
- design a training and awareness programme which encourages practical application to secure skills and future competences.

The programme will be designed around the operators, team members and managers concerned. It will be structured to maximize the contribution of each individual and to develop his or her skills to the limit of his or her capability.

Early equipment management

A further goal of TPM is to reduce equipment life cycle costs and maximize added value by improving:

- operability (ease of use)
- maintainability (ease of maintenance)
- intrinsic reliability
- customer-led product and service development
- life cycle cost prediction, feedback and control

In Japan over the last twenty years many hundreds of companies have applied the above principles to their operations. The Japan Institute of Plant Maintenance (JIPM) has carried out stringent audits of TPM achievement and continuity, resulting in the award of PM excellence certificates to successful companies.

The WCS International approach to TPM is to suitably modify, adapt and apply the Nakajima principles to aid communication, taking account of local cultural strengths and industry sector needs without corrupting these well-founded and well-proven original principles.

2

Assessing the true costs and benefits of TPM

The justification of expenditure is a rational management activity and, in an ideal world, all choices would be made directly by comparing cost and benefit. In reality, many important decisions are made on gut feeling because information is not available or the options are so complex that the benefits are uncertain.

One of the underlying strengths of TPM is its ability to reduce complexity and provide the route towards systematic decision making (see Figure 2.1). Early problem detection and resolution through self-managed teams also helps reduce the volume of matters requiring management attention – providing management with time to manage. The benefits can be significant (see Table 2.1).

Experience shows that such benefits are delivered progressively with wider involvement of personnel (see Figure 2.2). Despite such evidence, many continuous improvement programmes involve only a small percentage of the workforce in anything other than implementation activities.

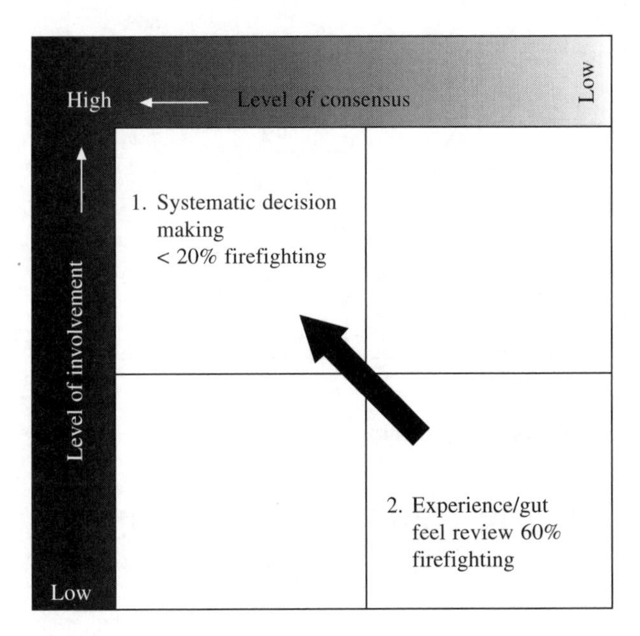

Figure 2.1 Potential decision styles

Table 2.1 World-class TPM plants

KPI plant	*Book covers*	*Clothing*	*Chemicals*	*Air conditioners*	*Lighting systems*
Breakdowns per months	18 5 3	707 $2^{1}/_{2}$ 15	200 4 10	250 6 5	387 4 33
Productivity	100 2 125	100 $2^{1}/_{2}$ 120	100 4 150	100 6 200	100 4 247
OEE (%)	55 3 75	71 3 120	100 4 160	65 6 88	71 4 88
Investment ($m)	1.9 5 3.4	1.0 per $ $2^{1}/_{2}$ 10.0	1.0 per $ 4 3.0	1.0 per $ 6 4.5	0.75 4 3.5

Key:

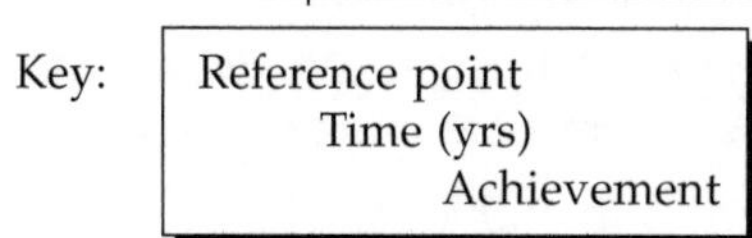

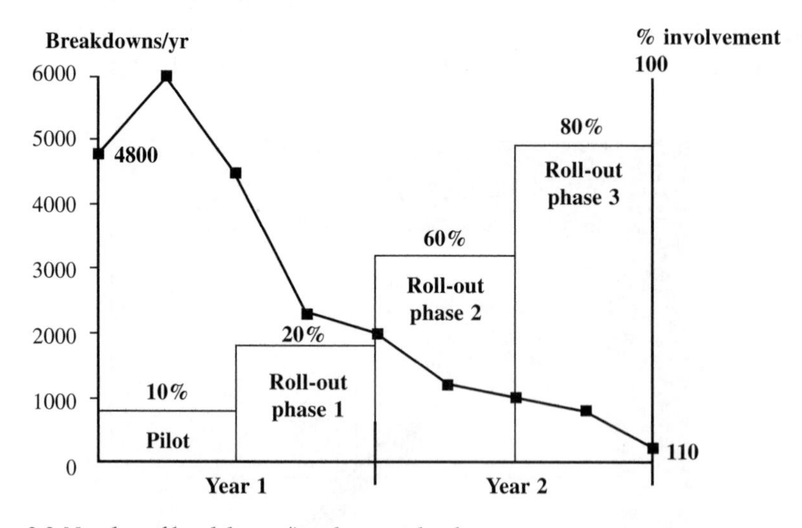

Figure 2.2 Number of breakdowns/involvement level

2.1 Challenging traditional thinking

Without the clarity which TPM brings, traditional management attempts to ignore the complexity of change by applying:

- a simplistic cost-down focus which inhibits learning and reinforces firefighting;

Table 2.2 Company-wide loss avoidance focus

Cost area	*Typical causes of hidden cost*		
	Availability	*Performance*	*Quality*
Material	■ Late receipt ■ Preparation losses	■ Material characteristic/ tolerance weaknesses ■ Planned waste	■ Manufacturing imperfections ■ Environmental/ customer usage conditions
Transformation	■ Supply failure/penalties ■ Disposal losses	■ Equipment idling ■ Tooling designs	■ Process deterioration ■ Human error
Equipment operating	■ Breakdown ■ Set-up and adjustment	■ Minor stops ■ Reduced speed losses	■ Rework ■ Start-up losses
Management	■ Lack of orders/resources ■ Planned shutdowns and unused capacity	■ Door-to-door losses ■ Non-added value activities	■ Logistics losses (not on time in full) ■ Equipment/ product management losses

- an internal rather than customer-focused perspective which stifles innovation;
- individual rather than team-based motivation which discourages idea sharing;
- political rather than effective management where presentation and image are as important as results.

This ignores the need for learning, assuming that one-off, quick-fix solutions are possible and that they are easily implemented.

The key to challenging this assumption is a clear understanding of how manufacturing costs really behave. The loss model provides such a picture, allowing management to make accurate predictions of the impact of strategic options.

Following on from the theme presented in Chapter 1, the model recognizes that a reduction in equipment loss is only part of what TPM can deliver. Management, material and transformation losses can be reduced, to transform operations in a way which touches all functions – making it an effective integrator of company-wide continuous improvement.

This opportunity will be missed with the simplistic cost-down focus or traditional management thinking.

2.2 The management challenge

Company-wide loss reduction provides the opportunity to:

(a) *either* produce the same in less time
(b) *or* produce more in the same time

The challenge for management is the need for clear strategies – in case (a) to secure a future with a smaller operation without slipping below critical mass, in case (b) to ensure larger market share or grow the existing one. The rational choice will almost always be (b), as this has the greatest potential to improve shareholder value and reduce unit costs. Taking on the entrepreneurial role is the real management challenge (see Figure 2.3).

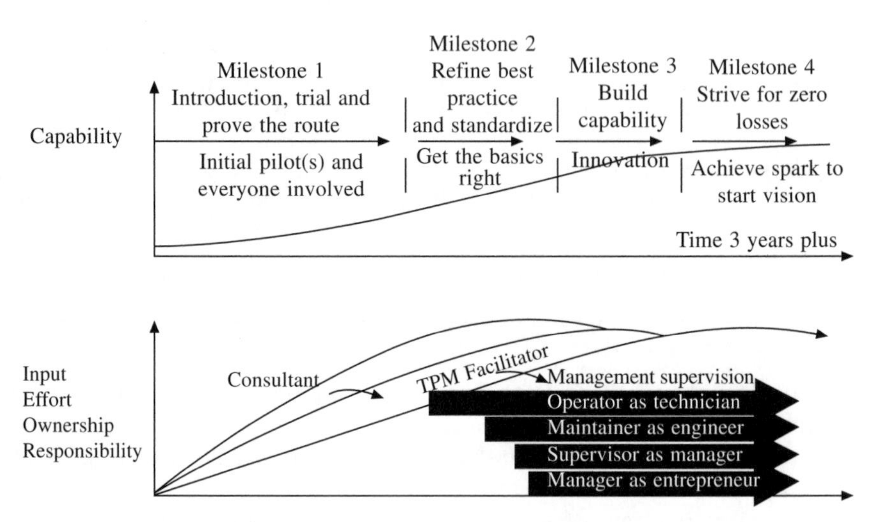

Figure 2.3 Developing capability

2.3 The potential to transform company performance

TPM provides a route to support either strategy by delivering:

- organizational activity which reflects the future needs of the changing customer/economic environment;
- an infrastructure to support team-based ways of working (management and shopfloor);
- decision processes based on systematic thinking to challenge accepted practices;
- team-based recognition and rewards systems which align short-term activities with long term-business goals;
- a focus on enhancing knowledge rather than defending your corner;
- improved OEE to enhance supplier/customer relationships;
- a mechanism to pull through improvements rather than pushing down initiatives;

- most importantly, it provides a development route to enhance the impact and develop the capability of key personnel (see Figure 2.3).

Repeated surveys of industry show that the limit to growth of manufacturing is not the lack of finance but the lack of both engineering and management skills. Undoubtedly, TPM's ability to release this is the greatest potential of all.

2.4 The loss model: a management development tool

In Chapter 1, we introduced the view that customer expectations should drive the company and therefore the company's operations' response (Figure 1.1). Then we introduced the management challenge presented by improved equipment effectiveness.

Here we look more closely at the TPM-derived loss model, an important management tool to deliver a Totally Productive Operation.

The example in Figure 2.4 raises some important issues concerning cost reduction and profitability:

- Producing 10 per cent more in the same time increases return on capital employed by 20 per cent.
- Producing the same in 10 per cent less time increases return on capital employed by 5 per cent (*reduced labour cost by 10 per cent).
- This highlights the importance of management focus on business development to create the environment for bottom-up continuous improvement.

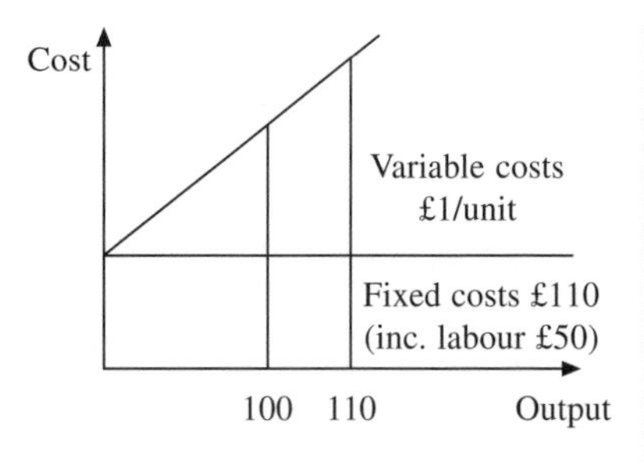

	OEE	*80%*	*Produce more 88%*	*Produce same 88%*	
(A)	Output	100	110	100	
(B)	Fixed Costs	110	110	105*	
(C)	Variable costs	100	110	100	
(D)	Total cost	210	220	205	
(E)	Unit cost D/A	£2.10	£2.00	£2.05	(D/A)
(F)	Contribution	1.05	1.15	1.10	
(G)	Unit sales price	3.15	3.15	3.15	
(H)	Total contribution	105	126.5	110	(F×A)
(I)	Return on capital employed	–	+20%	+5%	

Figure 2.4 OEE/Loss relationship

2.5 Meeting the challenge of change

Currency fluctuations, government policy and competitive pressure all ensure that the way we operate in five years' time will be different from the way we operate today. Reacting to new technology, new products and new legislation guarantees that every year will present a steep learning curve. Reflect on the last five years and expect twice as much change in the next five.

In this environment, the precise steps to deliver the chosen business strategy will evolve as opportunities present themselves. The ability to look forward and direct continuous improvement activities towards those opportunities is therefore vitally important.

3

The top-down and bottom-up realities of TPM

The reality of implementing TPM concerns two different dimensions: top-down strategic direction and bottom-up delivery of improvement. A strength of TPM is its ability to align both dimensions under a common goal.

This chapter looks at how apparently different motives can be integrated to the benefit of all stakeholders. This includes shareholders, licensing authorities, environmentalists as well as employees at all levels.

3.1 Setting and quantifying the TPM vision

Increasing pressure to drive down costs and eliminate waste in all its forms across the value/supply chain means the continuous improvement of our assets – both physical and people-related – is no longer an option. This also means that both the manufacturing and maintenance strategies, and their delivery, must fit and reflect the company's business drivers and strategy. It is customers who ultimately drive our business, and we therefore need to specify the necessary responses to satisfy and exceed these expectations by adding quality, performance and reliability – in all that we do.

Our own consultancy operation aspires to help manufacturing and process industry to realize its full potential in terms of customer service, cost, quality, safety and morale through the powerful enabling tool of TPM.

Determined world-class pacesetters will continue to use TPM as a key enabling tool to ensure a sustainable and profitable future for 2000 and beyond. TPM unlocks your installed productive capacity by unlocking the potential of your people, because Today People Matter!

The TPM loss model is a tool that predicts how costs will behave as a result of continuous improvement. This provides a feedforward mechanism, as opposed to 'feedback', to help management identify potential gains and direct priorities towards meeting and exceeding customer expectations.

What is a loss?

Each loss category is a legitimate top-down 'model' of a type of shopfloor problem, i.e. opportunity.

The use of loss categorization will be familiar to those who analyse equipment problems. For example, experience shows that for every breakdown there are around 30 minor stops and 300 contributory factors (Figure 3.1). Breakdowns are the result, not the cause or symptom.

With breakdowns, the contributory factors include scattering of dust and dirt, poor equipment condition and human error. Progressively reducing and eliminating these provides the organizational learning necessary to achieve zero breakdowns.

In addition to equipment losses, the loss model covers management, energy and material loss categories, providing a complete picture of operations' potential.

Building a loss model helps to create a top-down view of what the company might achieve by avoiding such losses (Figure 3.2). It also provides a basis for objectively comparing potential return on investment for improvement options. This is not an exact science, but it provides management with a continuous improvement framework for making sound management decisions (Figure 2.1).

The deployment route is provided via the pillar champions (Figures 3.3

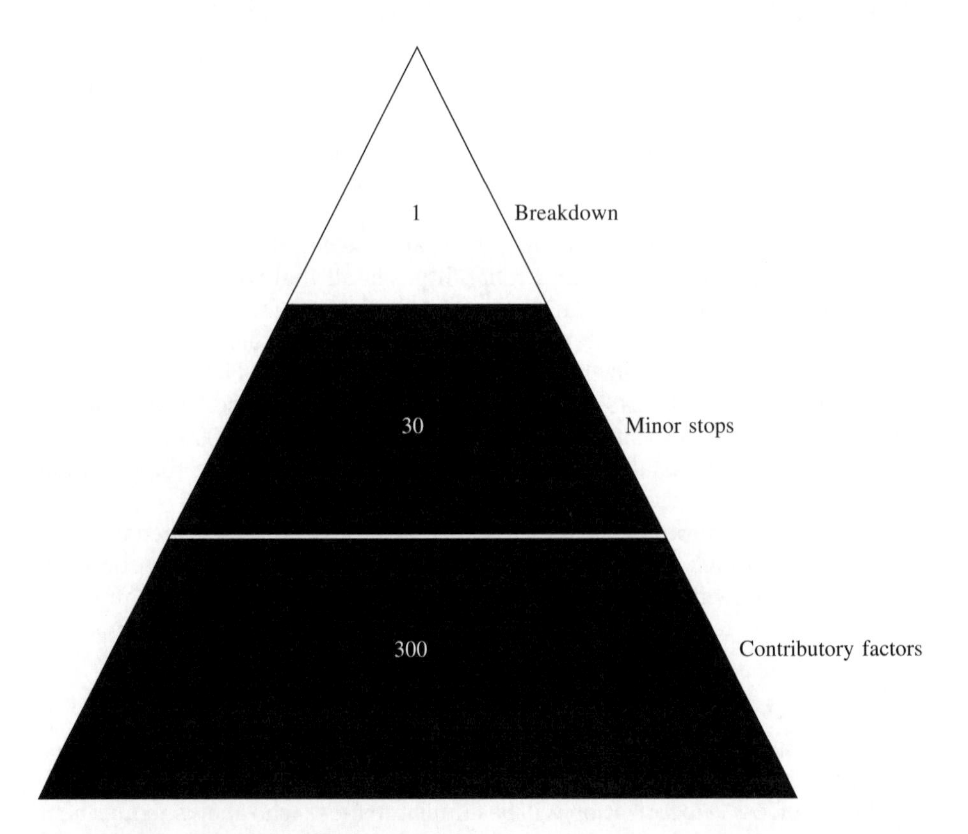

Figure 3.1 The structure of breakdown losses

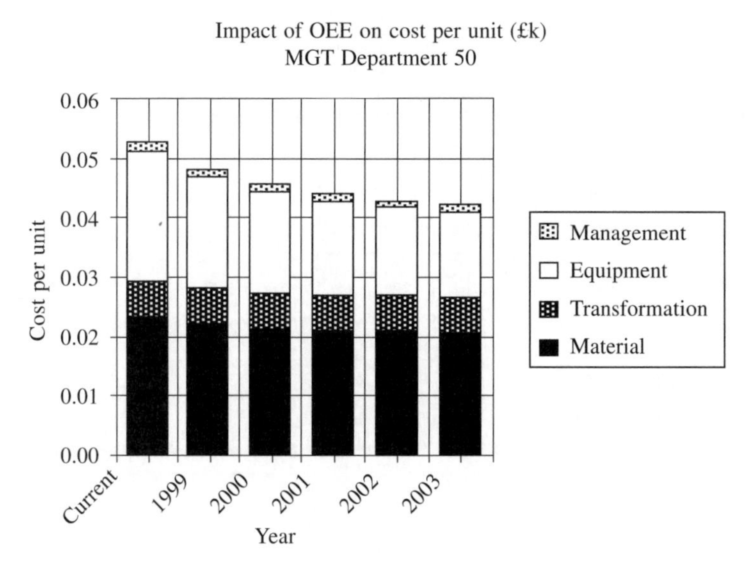

Figure 3.2 Loss deployment examples

Location: MGT
Version: 1.0 11 April 99
Top sheet attached [Y][N]
Scope of system: Department 50, A1, D2/3
Model potential gain: £568 500

Completed 4 | 1 Tactic deployed
Started 3 | 2 Tactic accepted and KPIs in place

No.	Tactic description	Cost	Forecast benefit	Resp.	Status
1	Improve OEE to release one miller for new business development (overhead reduction)	£?	£181 500	OEE	⊕
2	Standard shift working (contributes to 7) to raise lowest shift productivity to average	£?	£68 500	OAC	⊕
3	Standardize planned maintenance and carry out refurbishment to reduce sporadic losses by 25%	£?	£100 500	MAC	⊕
4	Refine/training in core competences to improve flexibility and reduce avoidable waiting time (contributes to 2)	£?	£150 000	SD/ LOG	⊕
5	Improve bottleneck resource scheduling to reduce avoidable waiting time by 50%	£?	£54 000	LOG	⊕
6	Reduce human intervention during equipment cycle to improve productivity (contributes to 1)	£?	£10 000	EEM	⊕
7	Improve best practice and technology to halve the quality failures	£?	£4000	EEM/ OAC	⊕

Champion responsibility: OEE = Overall equipment effectiveness. OAC = Operator asset care. MAC = Maintainer asset care. SD = Skill development. EEM = Early equipment management. LOG = Logistics

Figure 3.3 Pillar champion loss deployment target and activities

and 3.4). They are the custodians of the various TPM tools and techniques and through them the priorities and expectations of shopfloor teams are co-ordinated.

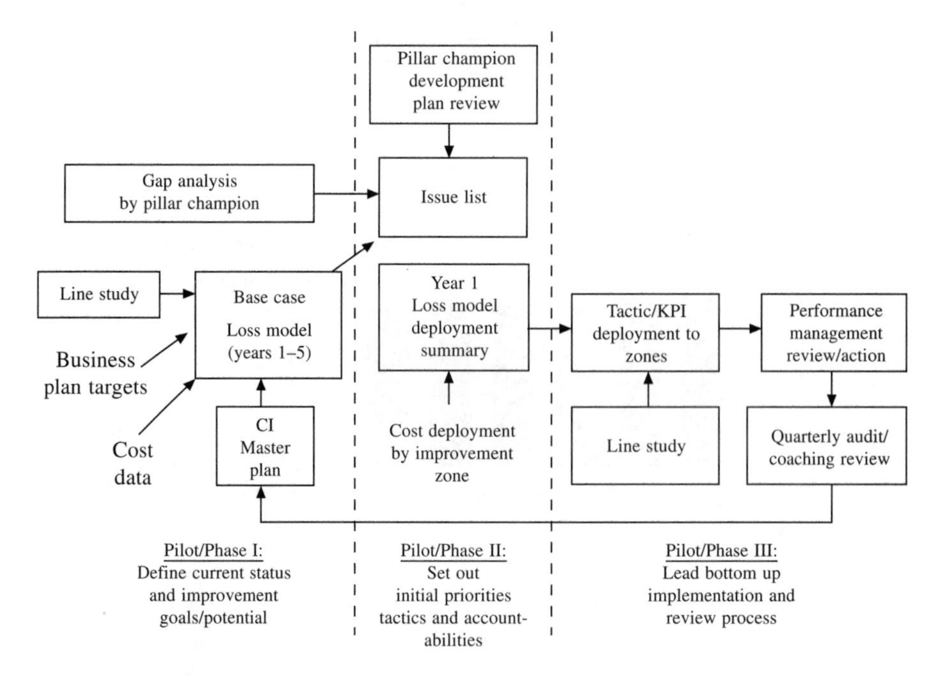

Figure 3.4 Top-down loss model development and deployment

Proactive management

Experience shows that the implementation of the loss model benefits from accurate data, the use of qualitative information, reinforced by data capture, is a powerful development tool for management in its journey from reactive to proactive management.

Linked to the results of the TPM audit/coaching process, each reiteration hones management's ability to *pull through* improvements rather than *push down* initiatives, making it an invaluable and integral part of the TPM methodology.

Typical cost/benefit profile

The cost/benefit profile in Figure 3.5 depicts the impact of introducing TPM into an organization and the effect this has on the OEE and, in turn, the overall maintenance budget.

The initial bow wave is the effect on the budget of the initial training, restoration/refurbishment and the time impact. Experience shows, however, that a 1 per cent improvement in the OEE is equivalent to between 5 per cent and 20 per cent of the annual maintenance budget.

So instead of simply attacking the tip of the iceberg, TPM flushes the 'hidden losses' to the surface for step-by-step elimination (Figure 1.3).

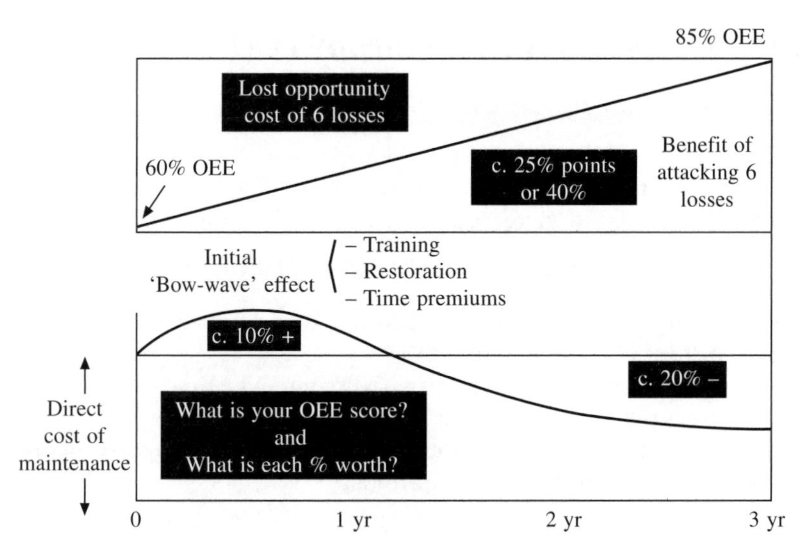

Figure 3.5 Typical cost/benefit profile

What can TPM give my business?

The benefits from TPM implementation are outlined under the following three categories and are shown in the case study examples that follow.

Business benefits

- Planning with confidence through the supply chain to provide what customers want, when they need it, just in time, right first time.
- Flexibility – being able to react quickly to market changes without high levels of stock.
- Improvement in OEE as a measurable route to increased profitability.

Equipment benefits

- Improved process capability, reliability, product quality and productivity.
- Economical use of equipment throughout its total service life starting from design, called TPM for Design or Early Equipment Management.
- Maximized efficiency of equipment.

People benefits

- Increased utilization of hand/operational skills, teamworking and problem-solving skills.
- Practical and effective example of teamworking, including TPM in Administration for the support functions.
- Trouble-free shifts, because value-adding activities become proactive rather than reactive.

Some examples of benefits from TPM

Case A: chemical processing plant

- By-product output constrained by capacity
- 5 per cent increase in OEE to 90 per cent
- Worth £400 000 in increased contribution per annum

Case B: manufacturing machining cell TPM pilot project

- OEE increased from 40 per cent to 72 per cent over six months
- Best of best OEE of 92 per cent
- 47 per cent reduction in set-up and changeover times
- 100 hours per month liberated via TPM improvements
- Additional manufacturing potential worth £48 000 per year by bringing subcontracted off-load work in-house

Case C: automotive manufacturer

- 15-year-old wheel balancer
- Average OEE before TPM = 45 per cent
- Cost of refurbishment = £8000
- Each 1 per cent improvement of OEE = £694 per annum
- OEE achieved after three months = 69 per cent
- Worth £17 000 per annum

Case D: polymer-based material producer

- Production line from raw material input to bulk reels
- Reference period OEE = 77 per cent
- Consistent achievement of best of best OEE = 82 per cent
- Value of achievement = £250 000 per annum in reduced costs
- One-off cost of improvements = £1000

Case E: cement plant

- Weigh feeder mechanism unreliability
- Reference period OEE = 71 per cent
- Best of best OEE achievement = 82 per cent
- Worth £35 000 in energy savings per annum
- Other TPM pilot improvements saved £300 000 per annum, plus avoidance of capital expenditure of £115 000

Case F: offshore oil platform with declining reservoir

After two years of using TPM principles:

- October 1997 achieved longest production run without shutdown since 1994
- Gas lift now at greater than 90 per cent efficiency compared to 40 per cent in 1995 and 60 per cent in 1996

- OEE reference period 60 per cent, current levels 75 per cent

Case G: pharmaceutical manufacturer

- TPM project actioned as part of a four-day facilitator training workshop
- Additional revenue generated worth £5 million per annum
- One-off cost of implementation £2000

TPM champion: 'I am glad we did not agree to a fee based on a percentage of the profits generated!'

Quotable quotes

'TPM is making rapid inroads into our reliability problems because of the structured approach which we have introduced. In the past, we have been shown the concepts, but we had to work out how to apply them. TPM is a much more practical and hands-on approach.'
Head of Continuous Improvement, European car manufacturer

'TPM is an excellent team-building process which helps develop the full potential of our people.'
Head of Maintenance

'Change initiated by the team through TPM is more rapidly accepted into the workplace than when imposed by management.'
TPM Champion

'If used effectively, TPM could be the most significant change to affect production and maintenance since Japan's entry into the car market.'
Manager, Continuous Improvement

'The main thing I've learnt is that TPM is not an option for us, it's a must.'
Plant Manager after attending four-day TPM workshop

'If you haven't got the time to do things right the first time, how are you going to find the time to put them right? Eventually TPM gives you the time to do things right the first time, every time.'
Offshore Maintenance Manager

'TPM is a new way of thinking, the cornerstone of which is the involvement of all our employees. The end result is a more efficient factory, a more challenged workforce and most importantly a reliable, high-quality service to our customers.'
Operations Director, packaging company

'The OEE ratio is the most practical measure I have seen'
Senior Manager

What does it take (cost) to deliver TPM benefits?

TPM uses an integrated set of techniques as shown in the three-cycle, nine-step schematic (Figure 3.6). The implementation of this analysis is applied stepwise within improvement zones to refine production and maintenance best practice through practical application (Figure 3.7).

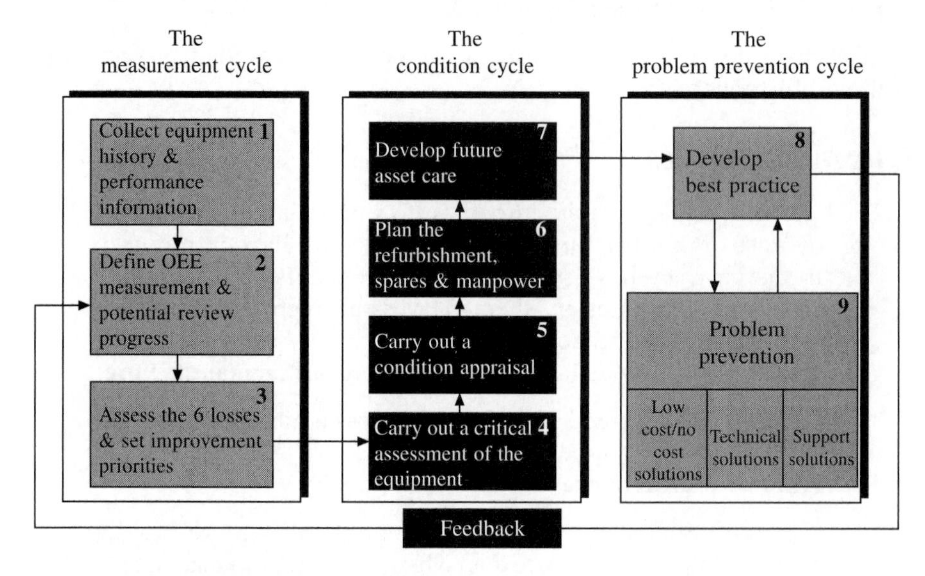

Figure 3.6 9-Step TPM improvement plan

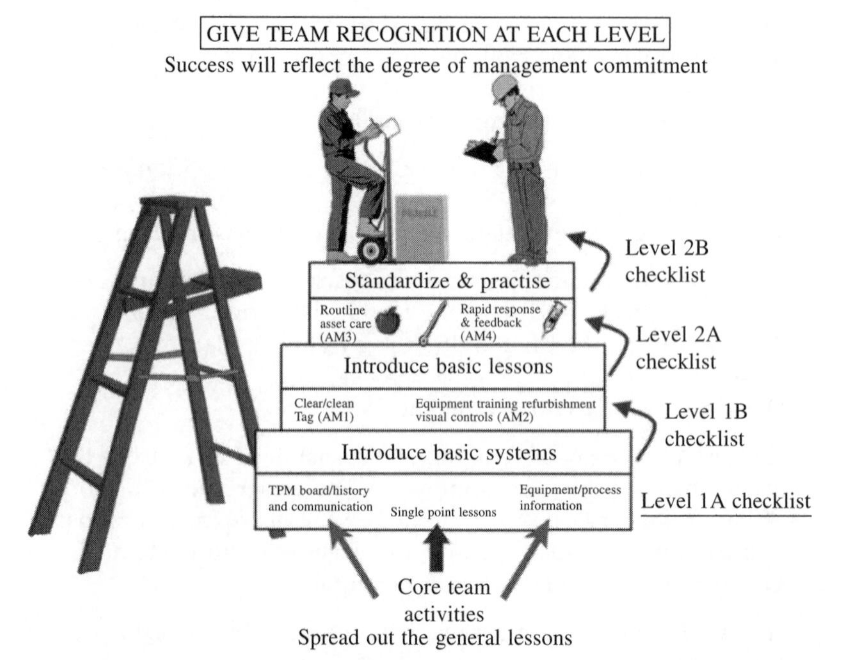

Figure 3.7 The TPM improvement zone partnership

Within these improvement zones, operators, maintainers and first line managers apply TPM by addressing cleanliness and workplace organization – eliminating dust, dirt and disarray. This is a positive step – cleaning is inspection, which is discovering abnormalities, which allows us to restore or refurbish, which gives a positive result, as well as a bright workplace, and ultimately gives our workforce back some self-esteem and pride of ownership. This is called the *5 Ss* or *CAN DO*.

Progress through these levels should be directed by asking:

- *Why* don't we know the true consequences of failure (both obvious and hidden)?
- *Why* does this part of the process not work as it is meant to?
- *Why* can't we improve the reliability?
- *Why* don't we have the skills to set the optimal conditions for the process?
- *Why* can't we maintain and progressively improve our technology to maintain those optimal conditions for longer?

The answer to all these questions is usually 'We don't know' because the shopfloor workforce have not been given the time, inclination and encouragement to find the answers. TPM gives the necessary time and motivation to do so. It also makes managers accountable for finding answers to each of those questions (i.e. pillar champions).

In summary, TPM recognizes that to achieve a reliable and flawless operation through continuous improvement, it is the people who make the difference. By unlocking your full productive capacity, TPM unlocks the potential of your workforce. You will need to invest around 5 per cent of your time to implement TPM and support continuous improvement. Like all good investments, this can be expected to provide a return on investment.

The nine-step TPM improvement plan is described in detail in later chapters and is at the heart of the practical application of TPM. It is a no-nonsense, no 'rocket science' practical application of common sense. The improvement zone implementation process is the way that this common sense becomes part of the routine. It takes time and tenacity, but the results are incredible.

Before moving into the necessary detail of the planning process and measurement of TPM, it is worthwhile to give an overview of TPM and to identify the key building blocks which will be explained in detail and illustrated by case studies in later chapters.

Whilst visiting Japan on a TPM study tour in 1992, we vividly remember being told by the Japanese Managing Director of a recognised world class manufacturer that

> '... in the 50s and 60s we had 'M' for Manufacturing. In the 70s we had 'IM' for Integrated Manufacturing. In the 80s we had 'CIM' for Computer Integrated Manufacturing'. He paused for a moment and then added '...For the remainder of this decade and 2000 and beyond, my company is going to be pursuing 'CHIM': Computer Human Integrated Manufacturing ... We have decided to re-introduce the human being into our workplace!'

Today, some eight years later our interpretation of that powerful message is that it certainly represents a challenge for all of us to develop and harness people's skills in parallel with advancing automation, as illustrated in Figure 3.8.

The challenge for many companies is to extend the useful life and efficiency of their manufacturing assets whilst containing operating costs to give a margin which will maximize value to their shareholders and, at the same time, offer enhanced continuity and security of employment. This statement is true whether the particular manufacturing assets are twenty years of age or are just about to be commissioned.

The more forward-thinking companies are linking this challenge to new beliefs and values which are centred on their employees through, for example:

- *Integrity* — Openness, trust and respect for all in dealing with any individual or organization
- *Teamwork* — Individuals working together with a common sense of purpose to achieve business objectives
- *Empowerment* — An environment where people are given both the authority and the resources to make sound decisions within established boundaries
- *Knowledge and skills* — Recognizing, valuing and developing the knowledge and skills of their people as a vital resource
- *Ownership* — A willingness on everyone's part to get involved and take responsibility for helping to meet the challenges of the future

Put another way, we can win the challenge by:

- working together
- winning together
- finishing first every time

This can be delivered by specific values, for example:

- *our people*
- working in a completely *safe* and *fit-for-purpose environment*
- where *quality* is paramount in everything we do
- where we have a *business understanding* linked to our activities
- and where *reliable equipment*, operated by *empowered* and *effective* teams, will ensure we finish *first every time*.

TPM, suitably tailored to the specific environment, can be a fundamental pillar and cornerstone to achieve the above goals, beliefs and values since:

- We all 'own' the plant and equipment.
- We are therefore responsible for its availability, reliability, condition and performance within a safe and fit-for-purpose environment.
- We will therefore ensure that our overall equipment effectiveness ranks as the best in the world.
- We will continuously strive to improve that world-class performance.

- We will therefore train, develop, motivate, encourage and equip our people to achieve these goals.
- We will therefore create an environment where our people *want* to challenge and change 'the way we do things here'.

The last statement is the fundamental future challenge for management if the previous statements are to mean anything in practice.

As the aerospace and nuclear power industries, with their relatively complex technologies and systems, emerged in the 1970s and 1980s, we had to respond with a selective and systematic approach. There developed *reliability-centred maintenance* (RCM), which considers the machine or system function and criticality and takes a selective approach, starting with the question: 'What are the consequences of failure of this item for the machine or system, both hidden and obvious?' For example, if the oil warning red light indicator comes on in your car, it is obvious that you are low on oil. The hidden consequence, if you do not stop immediately and top up the oil, is that the engine will seize! It is therefore good practice to check the oil level via the dipstick at regular levels. RCM takes a systematic approach, using appropriate run to failure, planned, preventive and condition-based strategies according to the consequences of failure.

TPM uses a similar logic, but emphasizes the people, measurement and problem-elimination parts of the equation and not systems alone. It emphasizes that people – operators, maintainers, equipment specifiers, designers and planners – must work as a team if they are to maximize the overall effectiveness of their equipment by actively seeking creative ways and solutions for eliminating waste due to equipment problems. That is, we must resolve equipment-related problems once and for all, and be able to measure that

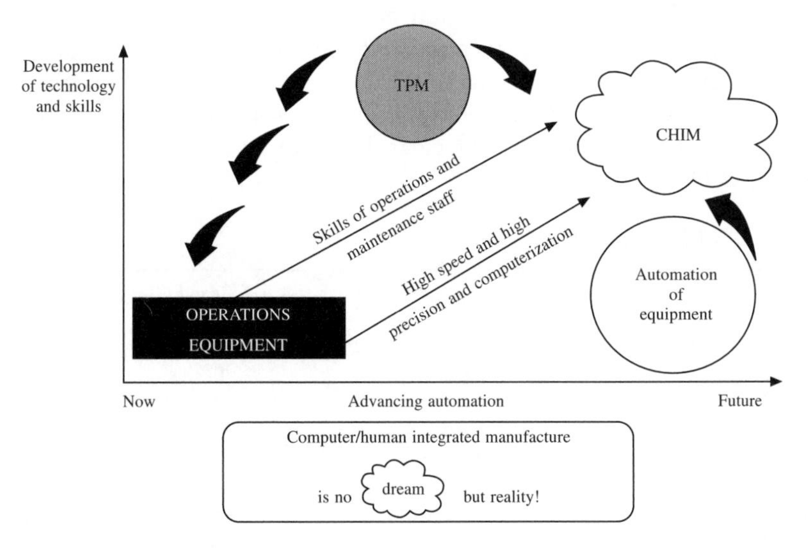

Figure 3.8 The future vision and impact on TPM

improvement. TPM is a practical application of total quality and empowerment working at the sharp end of the business – on your machines and processes.

It is useful to note here that the main trade union in the UK – the Amalgamated Engineering and Electrical Union (AEEU) – proactively pursues union/management partnership, where competitiveness is achieved *and* sustained '... through the knowledge worker in the knowledge-driven economy...'

Powerful stuff! And, furthermore, the AEEU sees the partnership benefits of TPM coming through the key points highlighted below in Figure 3.9.

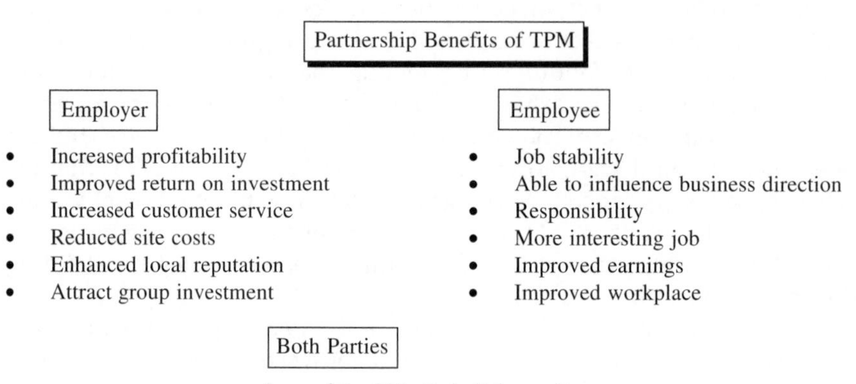

Figure 3.9 Partnership benefits of TPM

All sixteen bullet points are again powerful statements, but perhaps the final one of 'all seeking to continuously improve and add value' is the key.

Setting the vision is all very well, but we must also quantify the vision and make sure it reflects our business drivers and business objectives. In Table 3.1 below is a typical illustration of clear, hard targets for which TPM is the enabling tool.

Table 3.1 TPM-related targets

	Benchmark 1997	*Target 1998*	*Actual 1998*
All lines OEE	71%	90%	88%
Model line OEE	77%	90%	92.5%
B/D per month	387	40	33
Lead time days	45	15	15
Lost time accidents	1	0	0
Major set-ups	16 hrs	8 hrs	6 hrs
Minor stoppages per month	4650	1000	813
Reduction in product costs	100	90	91.5

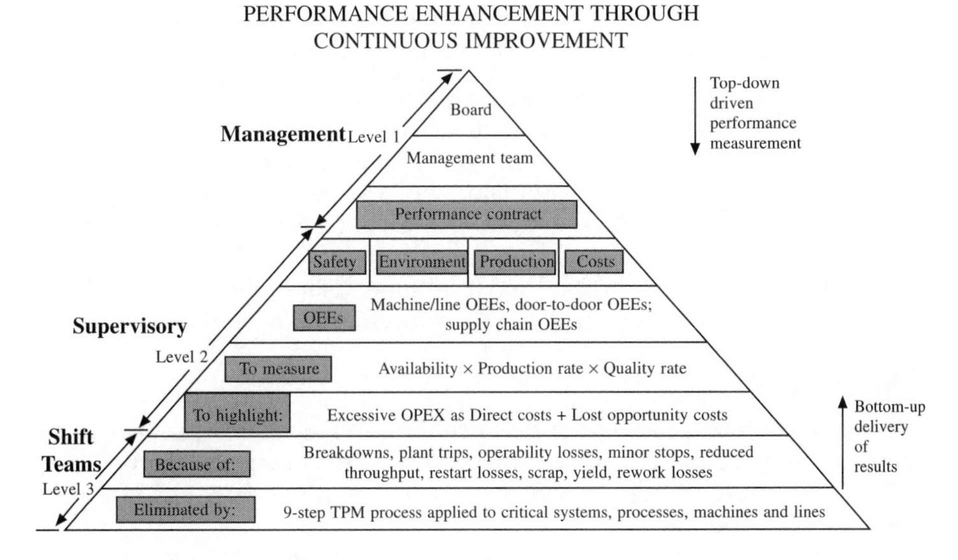

Figure 3.10 Top-down, bottom-up performance contract

In order to achieve the above targets, you need a 'top-down', 'bottom-up' delivery process as illustrated in Figure 3.10.

We will be expanding on this performance contract approach in later chapters. Suffice to say here that the TPM process can deliver the necessary business drivers, provided there is a logical and coherent performance triangle in place that reflects the organization structure which is aligned with clear roles and responsibilities.

From the 'bottom-up' perspective in TPM, management recognizes the simple fact that it is the operators of plant and equipment who are in the best position to know the condition of their equipment. Without their co-operation, no effective asset care programme can be developed and implemented. On the contrary, they can act as the senses (eyes, ears, nose, mouth, hands) of their maintenance colleagues, and as a team they can work out for themselves the best way of operating and looking after their machines, as well as resolving chronic equipment-related problems once and for all. They can also establish how to measure the resultant improvements.

TPM involves very little 'rocket science'; it is basically common sense. The problem is, it is quite a rarity to be asked to put our common sense to good use! TPM, however, does just that.

3.2 Analogies

In order to illustrate the application of TPM principles, three everyday analogies may prove helpful:

- the motor car (using the senses)
- the healthy body (defining core competences)
- the soccer team (creating the company-wide team)

Each is described below. At the end of the chapter there are two light-hearted stories. The first one is about an overhead projector operator and his maintenance colleague, which contains the best parts of the analogies in order to underpin the basically straightforward, but nonetheless fundamental, principles of TPM.

The second story relates to how a typical supervisor of the 'just do as I say' mould progressively changes to a 'let's work together to find the best way' style.

The motor car

A good analogy of using our senses, including common sense, is the way in which we look after our motor cars as a team effort between the operator (you, the owner and driver) and the maintainer (the garage maintenance mechanic) (see Figure 3.11).

As the operator of your motor car you take pride of ownership of this important asset. TPM strives to bring that sense of ownership and responsibility to the workplace. To extend the motor car analogy: when you, as the operator, take your car to the garage, the first thing the mechanic will seek is your view as to what is wrong with the car (your machine). He will know that you are best placed to act as his senses – ears, eyes, nose, mouth and common sense. If you say, 'Well, I'm not sure, but it smells of petrol and the engine is misfiring at 60 mph', he will probably say 'That's useful to know, but is there anything else you can tell me?' 'Yes,' you reply, 'I've cleaned the plugs and checked the plug gaps.' He won't be surprised that you carried out these basic checks, and certainly won't regard them as a mechanic-only job. 'Fine,' he might say, 'and that didn't cure the problem?' 'No,' you reply, 'so I adjusted the timing mechanism!' 'Serves you right then,' says the mechanic, 'and now it will cost you time and money for me to put it right.' In other words, in the final stage you, the operator, went beyond your level of competence and actually hindered the team effort. TPM is about getting a balanced team effort between operators and maintainers – both experts in their own right, but prepared to co-operate as a team.

As the operator of your car you know it makes sense to clean it – not because you are neurotic about having a clean car just for the sake of it, but rather because cleaning is inspection, which is spotting deterioration before it becomes catastrophic. The example in Figure 3.11 shows the power of this operator/ownership. In the routine car checks described, our senses of sight, hearing, touch and smell are used to detect signs which may have implications for inconvenience, safety, damage or the need for repairs or replacements. None of the 27 checks listed in the Figure requires a spanner or a screwdriver, but 17 of them have implications for safety. The analogy with TPM is clear: failure of the operator to be alert to his machine's condition can inhibit safety and lead to consequential damage, inconvenience, low productivity and high cost.

Routine checks:

✓	Tyre pressure:	extended life, safety	(eyes)
✓	Oil level:	not red light	(eyes)
•	Coolant level:	not red light	(eyes)
•	Battery:	not flat battery	(eyes)

Reasons: safety, consequential damage, inconvenience, low productivity, high cost

Cleaning the car: using our eyes

•	Spot of rust	✓	Steering drag	(touch, eyes)
•	Minor scratch	✓	Wheel bearing	(hear)
•	Minor dent	•	Clutch wear	(touch, hear)
✓	Tyres wearing unevenly	✓	Brake wear	(touch, hear)
•	Water in exhaust pipe	✓	Exhaust	(hear)
✓	Worn wipers	•	Engine misfire	(hear, touch)
•	Rubber perishing, trims	✓	Engine overheats	(smell)
✓	Oil leak	✓	Petrol leak	(smell)
✓	Suspension			

One operator to another at traffic lights

- • Exhaust smoke
- ✓ Front/rear lights
- ✓ Stop lights
- ✓ Indicators
- ✓ Door not shut
- ✓ Soft tyre

Message No spanner or screwdriver involved in any of the 27 condition checks
✓ Means check has safety implications (17 of 27)

Figure 3.11 Taking care of your car

We don't accept the status quo with our cars because ultimately this costs us money and is inconvenient when problems become major. In other words, we are highly conscious of changes in our cars' condition and performance using our senses. This is made easier for us by clear instruments and good access to parts which need regular attention. We need to bring this thinking into our workplace.

A healthy body

Figure 3.12 shows our second analogy, which is that healthy equipment is like a healthy body. It is also a team effort between the operator (you) and the maintainer (the doctor).

Looking after equipment falls into three main categories:

- *Cleaning and inspection* The daily prevention or apple a day, which prevents accelerated deterioration or wear and highlights changes in condition. The operator can do most, if not all, of these tasks where a technical judgement is not required.

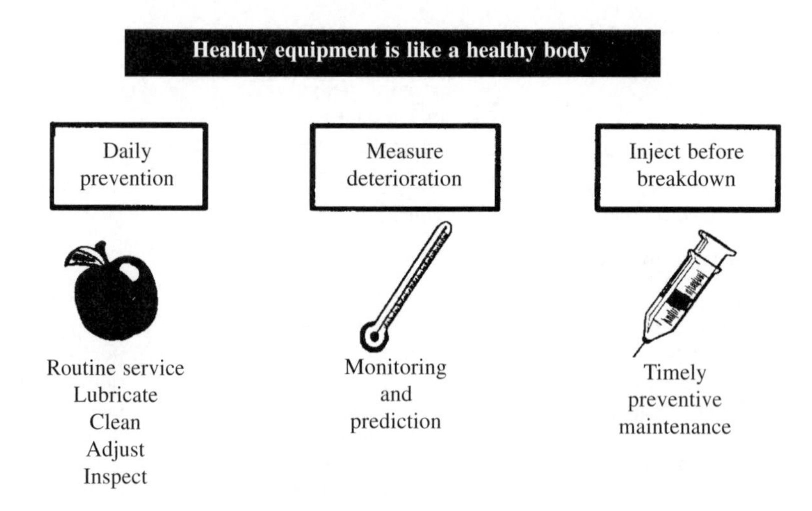

Figure 3.12 Healthy equipment is like a healthy body

- *Checks and monitoring* Measure deterioration or use the thermometer which highlights the trends or changes in performance. The operator can support the maintainer by acting as his ears, eyes, nose, mouth and common sense, thus allowing the maintainer to concentrate on the critical diagnostic tasks.
- *Planned, preventive maintenance and servicing* Inject before breakdown, which prevents failure by reacting to changes in condition and performance. The maintainer still does the majority of these tasks under TPM on the basis that a technical skill and/or judgement is required.

Perhaps the key difference in determining the asset care regime is that under TPM the operator and maintainer determine the routines under each of three categories. If you ask our opinion as an operator or maintainer and that opinion is embodied in the way we do things for the future, then we will stick with it because it is our idea. On the other hand, if you impose these routines from above, then we might tick a few boxes on a form, but we will not actually do anything.

The analogy works for activities across the company. Those who do the work are in the best position to define and refine core competences.

The soccer team

The third analogy emphasizes the absolutely critical aspect of teamwork. At every stage in the development of the TPM process, teamwork and total co-operation without jealousy and without suspicion are essential to success. In Chapter 7, we shall see how these teams are established and developed, but Figure 3.13 gives a pictorial representation of how the teams can function to maximum efficiency and minimum losses. Their job is to 'win', just as a soccer team on the field seeks to score and win the match. Just as the soccer team has the backing of a support group such as the coach, the physiotherapist, the manager and so on, the core TPM team also needs the proactive support of the designers, engineers, quality control, production control, union representatives and management.

In our soccer team the operators are the attackers or forwards, and the maintainers are the defenders. Of course, the maintainers can go forward and help the operators score a goal. Similarly, the operators can drop back in defence and help stop goals being scored against the team. They are both experts in their respective positions, but they are also willing to co-operate, help each other and be versatile. One thing is for sure in the modern world-class game: if we do not co-operate, we will certainly get relegated! The core team will invite functional help on the shopfloor when needed, and all concerned will give total co-operation with the single-minded objective of maximizing equipment effectiveness. Without co-operation and trust, the soccer team will not win. The core team on the pitch is only as good as the support it gets from the key contacts who are on the touchline – not up in the grandstand!

The TPM facilitator, or coach, is there to guide and to help the whole process work effectively. People are central to the approach used in TPM. We own the assets of the plant and we are therefore responsible for asset management and care. Operators, maintainers, equipment specifiers, designers

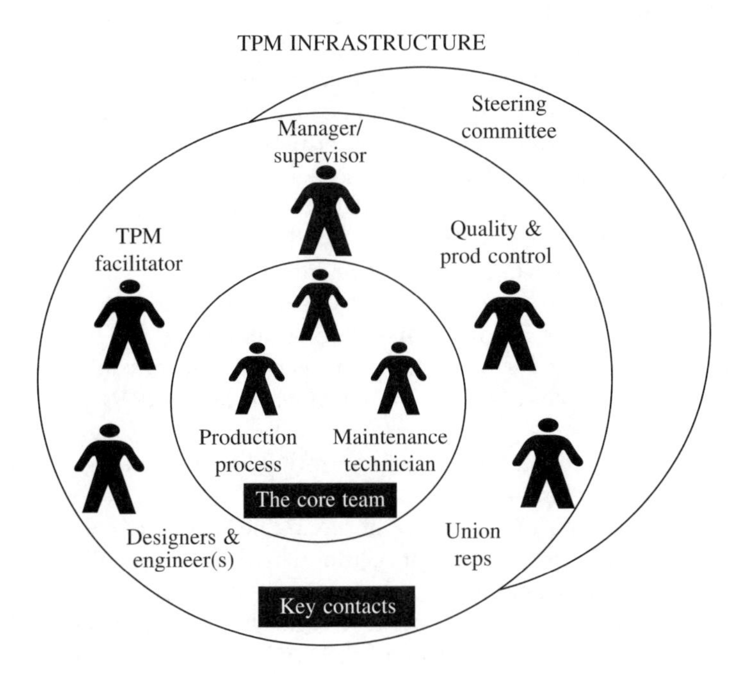

Figure 3.13 Teamworking

and planners must work as a team and actively seek creative solutions which will eliminate both waste and equipment-related quality problems once and for all!

An often-asked question is 'How does the 5S/CAN DO fit within the TPM activity?' One way of explaining it is to again use the analogy of the football team. As stated above, in this scenario, the operators are the attackers and the maintainers are the defenders.

However, they need a football to play with. The football is the structured and detailed nine-step TPM methodology of measurement, condition and problem-prevention activities, as applied to the critical machines and equipment (Figure 3.6).

There is, however, little point in having an excellent team with a powerful football if the pitch is in a dreadful state – namely the workplace and its organization. Under the TPM umbrella, the team takes responsibility for marking out the pitch, cutting the grass and putting up the goal nets and corner flags. This is the 5S/CAN DO activity which the team is responsible for, rather than it being delegated or subcontracted to a groundsman.

3.3 Overall equipment effectiveness versus the six big losses

The analogies above illustrate important common-sense dimensions of TPM philosophy. These combine to provide a powerful driver to improve OEE by reducing hidden losses.

In Figure 3.14 the tip of the iceberg represents the direct costs of maintenance. These are obvious and easy to measure because they appear on a budget and, unfortunately, suffer from some random reductions from time to time. This is a little like the overweight person who looks in the mirror, says he needs to lose weight and does so by cutting off his leg. It is a quick way of losing weight, but not a sensible one! Better to slim down at the waist and under the chin and become leaner and fitter as a result.

The indirect costs or lost opportunity costs of ineffective and inadequate maintenance tend to be harder to measure because they are less obvious at first sight – they are the hidden part of the iceberg. Yet they all work against and negate the principles of achieving world-class levels of overall equipment effectiveness.

In our iceberg example, the impact on profitability is in inverse proportion to the ease of measurement. Quite often we find that a 10 per cent reduction in the direct costs of maintenance (a commendable and worthwhile objective) is equivalent to a 1 per cent improvement in the overall effectiveness of equipment, which comes about from attacking the losses that currently lurk below the surface. Sometimes this is correctly referred to as the 'hidden factory' or 'cost of non-conformity'. The tip of the iceberg is about maintenance *efficiency*; the large part below the surface is about maintenance *effectiveness*. One is no

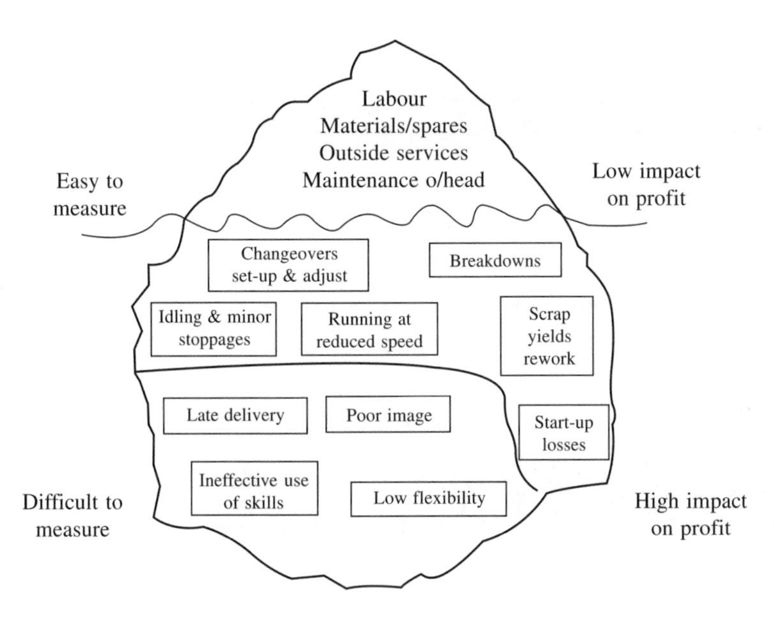

Figure 3.14 True cost of manufacturing: seven-eighths hidden

good without the other. Fortunately, the measurement of cost/benefit and value for money are central to the TPM philosophy.

In most manufacturing and process environments these indirect or lost opportunity costs include the following, which we call the *six big losses:*

- Breakdowns and unplanned plant shutdown losses
- Excessive set-ups, changeovers and adjustments (because 'we are not organized')
- Idling and minor stoppages (not breakdowns, but requiring the attention of the operator)
- Running at reduced speed (because the equipment 'is not quite right')
- Start-up losses (due to breakdowns and minor stoppages before the process stabilizes)
- Quality defects, scrap and rework (because the equipment 'is not quite right')

In Figure 3.15, we show the six big losses and how they impact on equipment effectiveness. The first two categories affect availability; the second two affect performance rate when running; and the final two affect the quality rate of the product. What is certain is that all six losses act against the achievement of a high overall equipment effectiveness.

In promoting the TPM equipment improvement activities you need to establish the OEE as the measure of improvement. The OEE formula is simple but effective:

$$\text{OEE} = \text{availability} \times \text{performance rate} \times \text{quality rate}$$

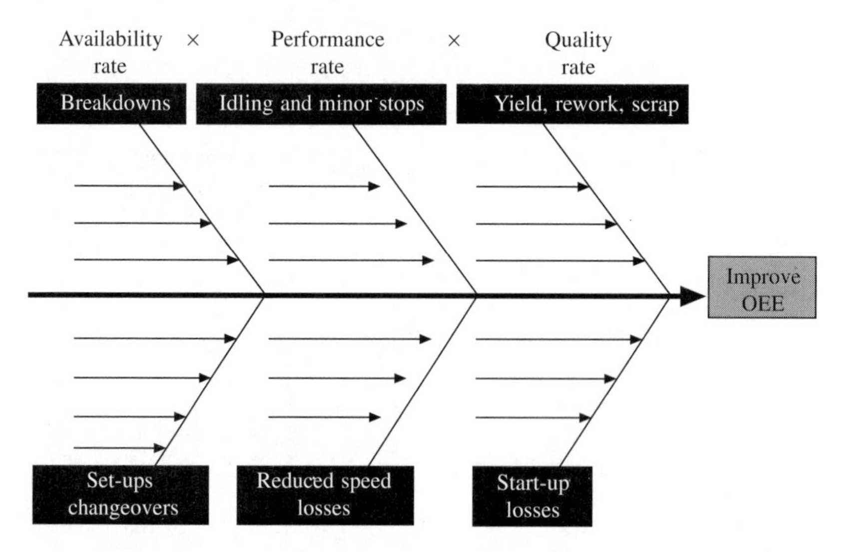

Figure 3.15 Six big losses

You will also need to determine your ultimate world-class goal or benchmark on the OEE measure. This should not be an idle dream: rather, it should be realistic, exacting, demanding and better than your competitors. Take each of the three elements in turn and set your ultimate goal. There may be a strong argument for your saying that the performance rate of the plant should be nothing less than 100 per cent, but be realistic. Figure 3.16 shows an example

BEST OF BEST OVERALL EQUIPMENT EFFECTIVENESS

Week No	OEE =	AV ×	PR ×	QR
1	83.8	90	94	99*
2	84.8	95*	92	97
3	80.9	86	96*	98
Average	83.2% =	90.3	94	98
B of B*	90.3% =	95	96	99

Question: What is stopping us achieving best of best consistently?

➡ **Answer:** We are not in control of the six big losses

➡ **However:** Best of best has a high belief level. Therefore teamwork and problem solving will lead to elimination of the six big losses.

Question: What is each percentage point improvement worth on OEE?

➡ **Answer:** 5 per cent improvement in OEE is often equivalent to 25 per cent of the annual maintenance budget

Figure 3.16 TPM Performance measure

which happens to give an average OEE of 83.2 per cent: your target may be higher.

Start to run the three measures, week by week, on your critical machines, lines and processes. Build up the notion of the 'best of best'. It is a very powerful and strong case. If we take the example shown in Figure 3.16, the best availability (week 2) × the best performance rate (week 3) × the best quality rate (week 1) gives an OEE of 90.3 per cent. What stops you achieving the best of best consistently? The answer is that you are not even in control of the six losses, far less eliminating them. This best of best, however, does have a high belief level: 'We have achieved it at least once in the last three weeks; the problem is we do not achieve each of the three OEE elements consistently.'

Each 1 per cent improvement on the OEE represents a significant contribution to profitability: it is the improvement below the tip of the iceberg. The vital issue is, of course, to determine what you can do with the improvement. Let us take a simple example:

	OEE	*Good units produced*	*Time taken*
Current	60%	1000	80 hours
Best of best	75%	1250	80 hours
		or 1000	64 hours

In the above example, consistent achievement of the best of best OEE from a 60 per cent base to 75 per cent is a 25 per cent real improvement. This means you can either produce 250 more units in the same time *or* the same number of units in 25 per cent less time – or, of course, some combination between these two levels.

The key point is that consistent improvement in the OEE gives the company and its management a *choice of flexibility* which they do not currently enjoy at the 60 per cent OEE level.

Table 3.2 presents most of the previous points as a summary of TPM's most desirable effects and the resultant benefits. TPM also gives us a clear vision, direction, involvement, empowerment and measurement tool for our future overall equipment effectiveness.

3.4 Getting started in your plant

As with most good practices, there is nothing particularly earth-shattering about TPM. The essence lies in the ability to focus the concepts and principles on the reality of the actual day-to-day situation. This means getting the climate right through front-line teamwork, aiming for motivation and ownership of the condition and productivity of equipment when it is up and running, rather than the 'I operate, you fix' traditional approach. This is easily said, but is potentially difficult to implement unless TPM is tailored to the specific

Table 3.2 TPM vision of the future

Feature	*Result*
• Machines run close to name-plate capacity	• Reduced capital expenditure need
• Ideas to improve often proposed by operators	• Ownership/success
• Breakdowns rare, and we achieve flawless operation	• Used to learn and teach the team
• Machines adapted to our need by our people	• Our machines will be better
• Operators and maintainers solve problems themselves	• Fewer delays and stoppages: enhanced self-esteem
• Cleanliness and pride in continuous improvement	• Good working environment
• More output potential from existing plant	• More profits and/or more control and choice

industry and local plant environment and business drivers (uptime versus downtime, output production, maintenance cost per unit of output, safety considerations, job flexibilities and so on) and, of course, the *essential* cultural and attitudinal perspectives.

Essentially we are talking about new ways of working, more effective and co-operative methods of carrying out essential asset care tasks and equipment-related problem resolution. This is achieved by improving the flexibility and interaction of maintenance and production, supported by excellent management, supervision, engineers and designers, plus systems, documentation, procedures, training, quality and team leading within an environment where safety is paramount.

TPM experiences in a wide range of industries confirm that it is essential to put handles on the issues before you can start to formulate a realistic programme of TPM-driven improvement with associated training, awareness and development. There is only one way to put handles on the issues, and that is to see and feel them at first hand. You *must* be prepared to spend sufficient time in the selected or proposed TPM plant so that you can see the reality and talk to the managers, superintendents, supervisors, engineers, designers, technicians, craftsmen and operators. As a result, you can understand where the plant is today and where it can realistically go for the future using the TPM approach. Whilst in the plant you can also formulate the training and awareness requirements as a properly thought-out plan with clearly identified benefits, costs, priorities, milestones, timescales, methods and resources. Each and every plant is like a thumbprint: it is unique and has to be treated as such.

3.5 TPM implementation route (Overview)

In helping our customers to introduce TPM principles, philosophy and practicalities into their company, we have developed a unique and structured step-by-step approach which is illustrated in Figure 3.17. It is a journey which comprises:

- securing management commitment;
- trialling and proving the TPM route as part of the policy development;
- deployment of that policy through four milestones, based on geographic improvement zones.

Typical timescales shown will, of course, vary according to the size of the operation, the amount of resource that is committed and the pace at which change can be initiated and absorbed. All these key questions, plus cost/benefit potential, are addressed within the scoping study or 'planning the plan' phase. Thorough planning is an essential forerunner for successful implementation.

Secure management commitment

Securing the necessary management commitment comprises three main elements:

- Senior management workshop
  The objectives of the management workshop are essentially to set and agree:

  - how TPM will fit with the business drivers and other initiatives;
  - how TPM needs to be 'positioned' for the site;
  - a management control system for the total programme;
  - the TPM vision for the company/site/plant.

- Plant-specific scoping study
  This describes the objectives of the plant, or site-specific scoping study, which is the essential tailoring of the implementation plan for the particular site.
  As indicated, this scoping study is carried out over a two to four week period, culminating in a local management review session to gain buy-in and commitment to the specific programme and as the launch-pad for implementation.
- Four-day hands-on workshop
  The final stage of securing the necessary management commitment is a four-day hands-on TPM workshop carried out on 'live' equipment in the host plant. The delegates will comprise a cross-section of senior management, potential TPM facilitators, union representatives (if appropriate) and some key operators and maintainers.

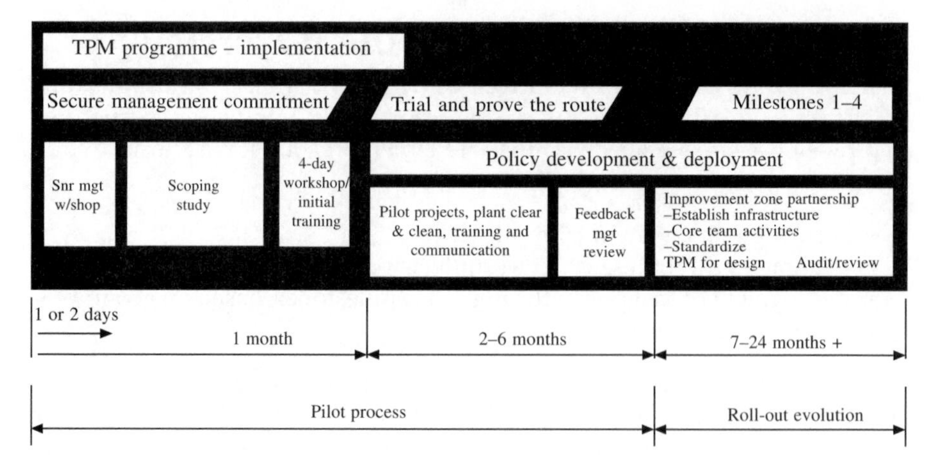

Figure 3.17 TPM implementation process

Trial and prove the route

This is the key phase for moving the TPM process from 'Strategic Intent' to 'Making it Happen', concentrating on focused improvements on the pilots using WCS's unique nine-step TPM improvement plan and getting everyone involved via the plant clear and clean activities of the 5S/CAN DO philosophy. This phase also includes setting up the TPM infrastructure, including the Steering Group, TPM facilitator and the TPM pillar champions.

Milestones 1 to 4 of the roll-out process

The roll-out builds on experience gained during the pilot process to ensure that the four development milestones of Introduction, Refine best practice and standardize, Build capability and Strive for zero losses become a *reality*, so that TPM becomes a 'way of life'.

In our experience it is vital to tailor your TPM implementation plan, not only to suit the differing cultures and industry types, but also to recognize the sensitivity of local plant-specific issues and conditions.

3.6 What is the 'on-the-job' reality of TPM?

TPM is different from other schemes. It is based on some fundamental but basically simple common-sense ideals:

1. We must restore equipment before we can improve effectiveness.
2. We can then pursue ideal conditions.
3. We must relentlessly eliminate all minor (as well as obvious major) defects, so that the 'six losses' are minimized, if not eliminated altogether.
4. We can start by addressing cleanliness – eliminating dust, dirt and

disarray. The philosophy here is that cleaning is checking, which is discovering abnormalities, which allows us to restore or improve abnormalities, which will give a positive effect, which will give pride in the workplace and will give our workforce back some self-esteem.

5. We should always lead the TPM process by asking 'why' five times (see page 31).

We usually don't know the answers to these questions because we have not been given the time, inclination and encouragement to find them. TPM gives us the necessary method and motivation to do so.

Most companies start their TPM journey by selecting a pilot area in a plant which can act as the focus and proving ground from which to cascade to other pilots and eventually across the whole plant or site.

Equipment losses often occur because the root cause of a problem is not eliminated (see Figure 3.18). When a defect occurs, production pressures and other constraints prevent a thorough investigation of the problem before solutions are applied. Instead, pit-stop 'quick fixes' are made which often result in performance and quality losses during operation. In many cases, defects which do not cause a breakdown are ignored and become part of the operating cycle of the equipment. Eventually, these defects recur and magnify, the same fixes are applied (under the same pressure) and the cycle continues. The TPM process breaks the cycle once and for all by identifying the root causes, eliminating them and putting in countermeasures to prevent recurrence.

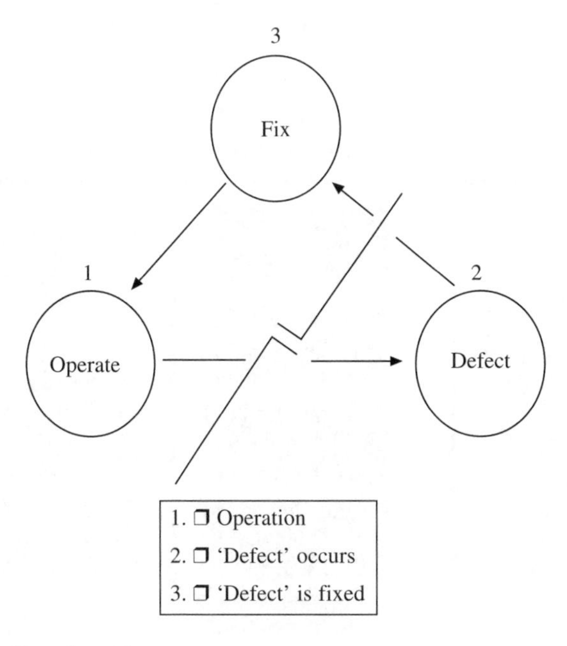

Figure 3.18 Breaking the cycle

TPM is about striving for:

- zero accidents
- zero breakdowns
- zero defects
- zero dust and dirt

3.7 What is the TPM improvement plan?

The TPM teams will follow the three-cycle, nine-step TPM improvement plan shown in Figure 3.19 in order to achieve and sustain world-class levels of overall equipment effectiveness by:

- measuring current equipment performance levels and setting priorities for improvement;
- developing a high level of knowledge of the equipment and its function;
- restoring the equipment to an acceptable condition in order to eliminate problems associated with deterioration;
- applying a level of asset care which sustains the new condition by exposing changes in condition and performance at an early stage and continuously improving techniques of prevention at the source of deterioration;
- eliminating problems using problem-solving and problem-prevention techniques to identify and remove the root causes;
- providing the best common practice and standardization of operation, equipment support and asset care for each piece of equipment across each shift, incorporating visual management and the training necessary to achieve this.

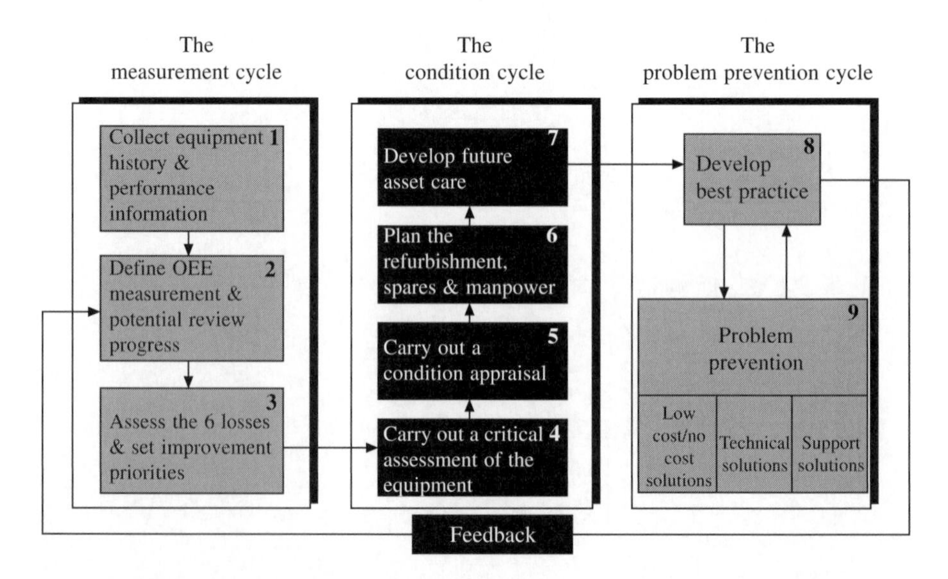

Figure 3.19 TPM improvement plan

In order to provide a precise and firm structure for the TPM process, WCS International has developed the nine-step TPM improvement plan, which has three distinct cycles:

1 The measurement cycle
2 The condition cycle
3 The problem prevention cycle

In order to ensure the quality of implementation for each TPM pilot application, the TPM team members are taken step by step through the main elements as shown in Figure 3.19.

Measurement cycle

Equipment history record

The TPM team analyse existing information sources and determine the future records to be kept with regard to the history of the equipment. This will aid future problem resolution.

OEE measurement and potential

In parallel with this exercise, the team carry out the initial measurement of overall equipment effectiveness in order to determine current levels of performance, the best of the best interim targets and the ultimate world-class levels.

Assess the six losses

This is a 'first cut' assessment of the impact of each of the six losses, and is usually aided by 'fishbone' analysis charts and 'brainstorming' in order to prioritize the losses.

Condition cycle

Critical assessment

In order to decide which are the most critical items, the TPM equipment team list the main sub-assets. Then they independently assess each of the sub-assets from their perspective and rank them on a scale of 1 (low) to 3 (high), taking into account criteria such as OEE, maintainability, reliability, impact on product quality, sensitivity to changeovers, knock-on effect, impact on throughput velocity, safety, environment and cost. The team should reach a consensus of ranking and weighting of the most critical items.

Other very useful outputs of the critical assessment are that it:

- helps to build teamwork;
- helps the team to fully understand the equipment;
- provides a checklist for the condition appraisal;
- provides a focus for future asset care;
- highlights safety-critical items;
- highlights weaknesses regarding operability, reliability and maintainability.

Condition appraisal

Following the above step, the TPM pilot teams can start the equipment condition appraisal. Typically a team will comprise a team leader with two operators and two maintainers as team members plus, of course, the TPM facilitator. The TPM pilot team should start the condition appraisal by carrying out a comprehensive clean-up of the equipment to see where deterioration is occurring. This will include removing panels, so that a deep clean can be carried out.

Refurbishment

The next task of the TPM team is to decide what refurbishment programme is required in order to restore the equipment to an acceptable level of condition from which the ideal condition can be pursued. Depending on the extent of refurbishment needed, up to three work packages may be appropriate:

- Work that can be done 'on the run'
- Work requiring an 8- to 24-hour outage
- Project work involving redesign and/or subcontractors

Future asset care

Whilst completing the condition appraisal the team can also determine the future asset care programme in terms of who does what and when. They can decide the daily prevention routines – the lubricate, clean, adjust and inspect activities, which will be carried out by operations staff. They can also decide upon the condition monitoring activities needed to measure deterioration – remember that the best condition monitor is the operator using the machine. The operator acts as the ears, eyes, nose, mouth and common sense of his maintenance colleague and can call him in when things start to go wrong and before they become catastrophic. Finally, the team also decides upon the regular PPM – the planned, preventive maintenance – and contribute to the condition monitoring, whilst the maintainer does the PPM scheduled work. This asset care step also determines the spares policy for the specific equipment under review.

Problem prevention cycle

Best practice routines (BPRs)

The TPM pilot team will develop its own BPRs regarding the equipment operation and asset care policy and practice. All these feed back into an improved OEE score which will encourage the continuous improvement 'habit' – this is central to the TPM philosophy. As in total quality, the personnel will also become empowered!

Problem prevention

This final step is about getting at root causes and progressively eliminating them. *P–M analysis* is a problem-solving approach to improving equipment effectiveness which states: There are *phenomena* which are *physical*, which cause *problems* which can be *prevented* (the four Ps) because they are to do with *materials, machines, mechanisms* and *manpower* (the four Ms).

This is the acid test of the TPM pilot(s) since the teams are trained, encouraged and motivated to resolve (once and for all) the six losses which work against the achievement of world-class levels of overall equipment effectiveness. These problem-solving opportunities can usually be classified as:

- operational problems/improvements involving no cost or low cost and low risk solutions;
- technical problems/improvements, often involving the key contacts and also some cost and, hence, risk;
- support services and/or support equipment problems/improvements which can involve the key contacts and some low cost but low risk.

Obviously the elimination of problems needs to be developed into the best practice routines, the impact of which will feed back into an improved OEE figure.

In order to summarize the previous explanation, Figure 3.20 shows that TPM involves a team of craftsmen and operators who are supported by their key contacts and who follow the TPM improvement plan through initial

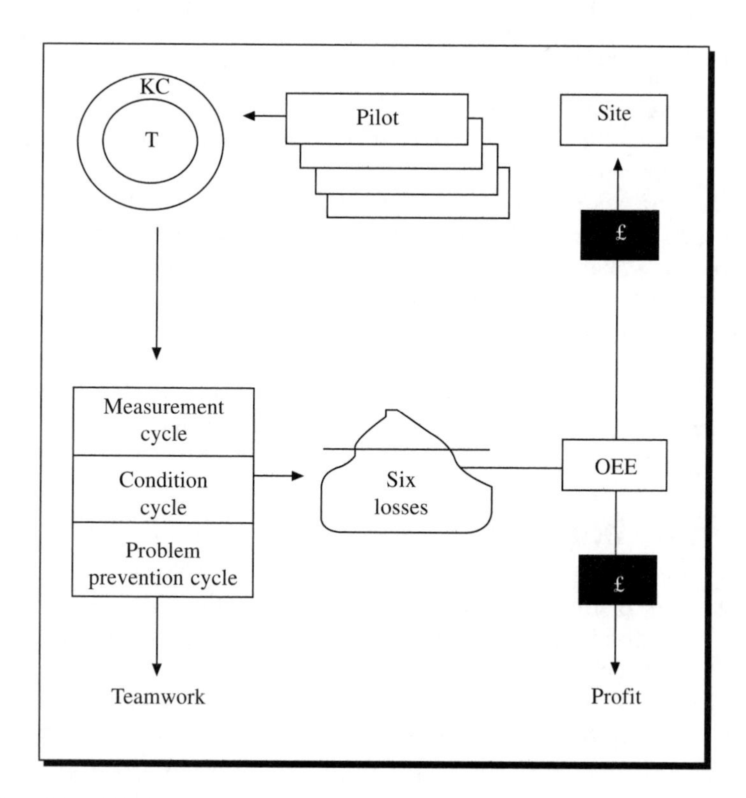

Figure 3.20 The TPM journey

pilots in order to eliminate the six major losses. Their progress is measured by improvements in the overall equipment effectiveness which allows the team to understand the need to continuously improve. Finally, the TPM process will only work provided it has the sustained commitment of everybody – which, of course, must start from the top.

3.8 Some TPM guidelines to follow faithfully

When assessing the potential benefits of TPM, *always* consider:

- the hidden benefits of being able to produce, say, 15 per cent more with the same resources, within a year;
- the difference it would make to your job if things happened as planned;
- what you want your company to be like in three to five years' time, and the impact that TPM would have on achieving the vision;
- that unless you are prepared to change and direct resources towards continuous improvement, you will fail. There is no other investment like it;
- you cannot afford to carry passengers. Everyone must be involved.

When assessing the potential benefits of TPM, *never*:

- look only at direct costs;
- believe you are achieving TPM effectively already, unless you can show consistent benefits;
- look for a short-term fix. Expect gains, but you will not keep them unless TPM is part of a long-term commitment to strive for zero losses;
- expect the change to be limited to the shopfloor. All departments have an influence on equipment effectiveness;
- expect it to be easy. TPM will need a high degree of motivation and determination to succeed. Your company will need to learn how to work together to succeed. Most of the resistance will come from management. It will be passive and difficult to detect.

3.9 Final thoughts

Adapting the principles of TPM to suit our differing cultures is one thing – tailoring them to suit your specific industry is another. The most vital issue is to recognize and incorporate the local plant-specific needs into TPM-driven improvement processes. TPM is not a programme or project with a start and a finish but, on the contrary, is a continuous improvement process, so it becomes a key part of 'the way we do things here'. As mentioned at the beginning of the chapter, below are two perspectives on the impact of TPM on 'the way we do things here'.

3.10 The overhead projector analogy

Good morning, everybody, my name is Peter Willmott and my job is to operate this overhead projector (Figure 3.21). I have worked for the OHP Company for twenty years now and, provided I have walked in vertically every day and have been warm to the touch, nobody in management has taken too much notice of me!

Times are a-changing, however, and seemingly for the better. Apparently, our Managing Director has been to visit one of our competitors and has seen how they look after their equipment and, perhaps more importantly, their people. They practise a thing called Total Productive Maintenance or TPM. Now that it has been explained to us, we prefer to think of TPM as Total Productive Manufacturing, as it's about teamwork between production and maintenance. It's quite simple really. Common sense, you might say, and the

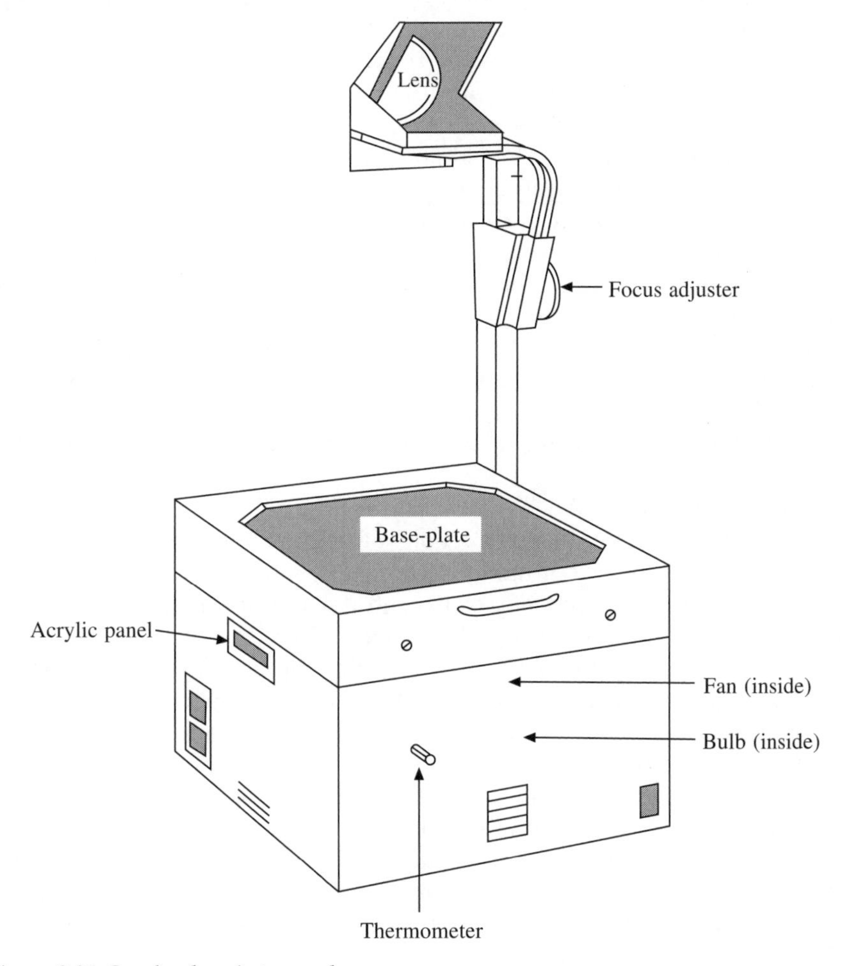

Figure 3.21 Overhead projector analogy

best part about it is that it involves no rocket science; but it does involve me – the operator of this overhead projector – and my maintenance colleague here, Joe Wrench. I have known Joe for ten years and he has always been good at fixing things. In fact, we have jokingly referred to him as 'Joe'll Fix-It' as he works for the 'GITAFI regime': get in there and fix it! When I first joined the company, I remember seeing Joe leaning against a pillar near my machine one day, and I asked him what his job was. He replied that he was a 'coiled spring, waiting to spring into action'!

As I say, things are changing for the better, and Joe and I are encouraged to work as a team as far as operating and looking after this overhead projector is concerned. For the first time in twenty years, I have actually been asked my opinion about the equipment together with Joe's ideas, and we have come up with some good ideas. Let me explain them to you.

For a start, you will not thank me if I project this visual aid – the product – onto the ceiling, or if it is completely out of focus. I have actually been given a comprehensive training session on the correct operation of the overhead projector and, in fact, Joe and I have drawn up a simple ten-step start-up, operation and shutdown procedure for the OHP as a series of ten single-point lessons (SPLs) which are very easy to follow and highly graphic and colourful to make the SPLs interesting.

Because we are being encouraged to look after the OHP and are given the time and support equipment to do it, I actually clean the lens and the projector base-plate at the start of each shift since it improves the quality of the product – in this case the presentation of the visual aid – and I make sure I adjust and focus it correctly before starting the shift. By the way, I also make sure I cover up the base-plate of the OHP at the end of the shift, as it can easily get scratched and damaged if I do not do this simple chore. A new base-plate for this OHP costs £55.00, which is about 15 per cent of the cost of a complete new OHP. It is also inconvenient, as it takes about three hours to change over the old one for a new one.

Anyway, as I said, Joe Wrench, my maintenance colleague, and I have been given the training, time and encouragement to sort out the best way of running this piece of equipment. Let me tell you what we have decided to do. Not, you will notice, some clever chap from central planning, but Joe and I. We are in a team now, and Fred Whitlock, the ex-supervisor, is now our Team Leader. (You can read his story about TPM later in this chapter.) Since he has been on a TPM facilitator course he's changed for the better: he asks our opinion about things and he actually takes the time out to come down here to listen and discuss better ways of doing things with Joe and me.

One of the problems with this OHP is that the focus adjuster on this vertical arm here seems to wear out quite often, and the ratchet won't hold the lens head in focus. If this happens during the shift, Joe and I have decided that we don't need to actually stop the shift for a major repair. Instead, I can pin the ratchet with this wedge as a temporary measure whilst I complete the shift. We can, in effect, run it to failure, and the only thing I make sure I do is to let Joe know that he will need to change the ratchet focus adjuster as soon as he's got time to do it. By the way, Joe and I have put forward a proposal to production engineering to use a closer tolerance and age-hardened ratchet so that this problem is resolved once and for all. This will mean our spares costs will go down and I won't have to mess about jamming a wedge in here as a temporary measure: botching up is a thing of the past. After all, if the handbrake ratchet on your motor car kept failing, you wouldn't put up with it, would you?

The other thing which Joe and I have discussed is the bulb-changing task on this OHP. I used to think the bulb was the most critical part of this machine, but it isn't. It's important, but not as critical as something else which I will tell you about later. Anyway, back to the bulb. They do fail now and again, and in the bad old days when the bulb went I used to switch off the machine and go for a cup of tea and wait until Joe got round to getting a new one out of stores – which is about half a mile away. Joe would then change it over and we would eventually get going again. I reckon we used to lose something like four hours a month on this 'breakdown' if the bulb went. Not any more, though, because Joe and I have thought about this problem as well. In fact, I've been on a half-day in-company bulb-changing course and I am now a fully accredited bulb changer! ... I'm certainly no electrician, but I am proud of what I've achieved.

What happens now if a bulb goes? Quite simple: I switch off the on/off switch here and walk over to the power point. I switch that off. I pull out the plug and bring it back here with the lead, so there is absolutely no way I can electrocute myself. I then remove the lid, take out the old bulb and put it in the waste-bin here; I do not leave it lying around as a future accident risk. Then, using a cloth, I take out the new bulb from its packaging; I use a cloth because it's a halogen bulb and if I get my sweaty hands on it, it will be useless. I then insert it here, replace the lid, take the plug and lead to the socket and re-energize the circuit. Switch on the on/off switch at the OHP and 'bingo' – we're back in production. I feel really good that I can change bulbs. I feel a better person all round.

There are some other important points about bulb changing which Joe and I have agreed. We keep two spare bulbs here by the machine, on this 'shadow board' – not 800 metres away in central stores. I always – without fail – record the fact that I've used a bulb so that Joe and I can build up an equipment history file on this OHP, so that we both have access to past problems – non-standard events, if you like – which will help us in our problem resolution sessions. At the moment, Joe and I are looking into the possibility of bulbs with a different power rating, as these current ones seem to be unreliable. Joe also thinks it may be something to do with dust and dirt ingress, but more on that later.

Now to the best part of this equipment care procedure which Joe and I have built up and which we are both pretty proud of. The most critical part of this OHP – given that we have a power supply, of course! – is the fan. For years, I'd always thought the bulb was the most important part. I didn't even realize there was a fan in the machine, far less understand that it's there to create an air flow across the bulb to keep it cool and so stop it overheating. If the fan goes, the bulb will most certainly blow, and my product, the overhead view foil, will probably melt in the process! I hadn't really thought about this before – mainly because I hadn't been asked to think about it!

It's quite interesting, really, because when I started to clean the OHP I could tell it was getting hot – or overheating, to be precise (because I like to be precise nowadays). Cleaning is inspection, and I'm really acting as the ears (sound), eyes (sight), nose (smell), mouth (taste), hands (touch, heat, vibration) and *common sense* of my maintenance colleague, Joe Wrench. And I don't usually need a spanner or screwdriver to use any of my God-given senses! Incidentally, I've learnt that common sense is in fact quite uncommon unless we're encouraged to use it! Anyway, back to the fan, the most critical part of my machine, the OHP. If, when I'm cleaning it, I notice it's getting too hot, or if it starts to make a noise or vibrates, I do one thing, and one thing only. I switch off the on/off switch, I pull out the plug and bring it back here to the machine, and then I get Joe to come and see what's wrong. It's beyond my level of competence or skill at the moment to go messing about with the fan, but I can and do act as the early warning system for Joe.

In fact, Joe and I have thought a lot about the fan and we are getting a bit more scientific about the early warning system – or condition-based maintenance or monitoring (CBM) as we call it. Rather than trust my 'feeling the heat' or 'hearing the noise' senses we've decided to drill a hole here in this precise position and we've inserted a thermometer with a red mark on the 40°C point. So during the shift I do three readings: after one hour, after four hours, and just before the end of the eight-hour shift. I can trend the readings and I keep them up on this visible wall chart so that both Joe and I can see the temperature trends, alongside the major event fault trends. Joe and I have made two other improvements as well. In fact, we're quite proud of these equipment improvements that we've implemented. The fan drive belts used to break quite often, so I've suggested we cut out a 100 mm × 100 mm panel on the side here and put an acrylic cover in place of the metal sheet we removed, so that we can look inside the OHP base and see if the belt is fraying before it actually breaks. It's simple really, and, we think, quite effective. In fact, all 80 of our other OHP machines are now fitted with the thermometer and the acrylic cover modifications.

The really exciting bit about this TPM process is when we get to Step 9, the problem prevention bit – where we need the 5 whys technique to get at the root cause and prevention routine as to why the bulbs were blowing or failing. The 5 whys process goes like this:

1 Why is the bulb blowing? Because it's overheating.
2 Why is it overheating? Because the air flow is insufficient.
3 Why is the air flow insufficient? Because the filter is blocked.
4 Why is the filter blocked? Because nobody cleans it.
5 Why isn't it cleaned? Because we didn't appreciate the importance of daily asset care – apple-a-day routines!!

Joe and I have developed best practice routines for this OHP, divided into three main areas:

- The 'apple a day' routines which I do as a matter of habit, such as cleaning and changing over the filter
- The 'thermometer' or condition monitoring routines which Joe and I share
- The 'injection needle' or planned maintenance which is still carried out by Joe where a technical judgement is needed.

The simple act of me cleaning the filter once a week and changing it every month means that we have extended the useful life of the bulb and the fan because it's not put under stress to drag the air in – and I get a 'hassle-free' shift. Joe and I are not content, however, with extending the life of the components – we are now looking for the source of the dust and dirt that gets into the filter in the first place. Most of it is because we are not looking after our general workplace areas. This is where the 5S/CAN DO activity has come into its own.

The point is that we, not someone from on high, have decided the best practice routines to operate and take care of our asset – the OHP. Also, we have decided who actually carries out each asset care task, how we carry it out, with what frequency and with what support tools and equipment.

It's our ideas, it's our disciplines that are important: we've got ownership and we work as a team. We've been given the time, the responsibility and the necessary training and encouragement to take ownership, and we like it. It's given us back some self-esteem. It's for maintenance to be productive, whoever does it!

Finally, Joe and I had our photographs put up in the reception area in the front office last month as recipients of the TPM team of the month award. Silly, really, but Joe and I felt quite good about it. Even my wife says I'm warm to the touch now!

3.11 The first line manager's view of TPM

(with thanks to Graham Davies of WCS International for his insights as an ex-Plant Manager and Supervisor)

The scenario

The principles of TPM and their adoption may be accepted by the upper management of any company, but the area which can make or break any real commitment to TPM is the first line management, usually called the Team Leader, who is often an ex-shopfloor person. Attitudes and hence 'buy-in' on the shopfloor are dictated more by the Supervisor than the Plant Manager.

As far as an operator or maintainer is concerned, the only person they relate to on a daily basis is their Supervisor. The majority of changes being implemented by upper management do not make a great deal of difference to them. The only concern in their mind is whether or not their Supervisor will put these changes into practice and how he sells it to them. If he is autocratic, then the opinion the operator and maintainer have of the company will reflect that; if he is an open-minded person who people respect and he takes their ideas on board, then they will have a different view of how the company operates.

My name is Fred Whitlock and, as a Team Leader, it is my job to implement the changes which management see fit to promote within our organization. I also feel, however, that with the experience I have, I should ensure that I interpret changes in a way that will be of benefit to our company and also make it easy for me to sell to the workforce.

I have worked for this company, man and boy, for thirty-five years. At fifty-two, I am at the point of knowing this place inside out and have the respect of my peers as someone who can sort out the problems that occur on my shift and within the department as a whole.

In the time that I have been working here I have seen all sorts of so-called 'initiatives'. I call them 'flavour of the month' because that's about as long as they last, and here they are again with another one. Yes, it's TPM, Total Productive Manufacturing – the best thing since sliced bread!!

The denial phase

I only have eight years left before I retire, and all I want to do is to have an easy life until then. I do not see why I need to take on extra responsibilities when I do not get enough time to do what I should be doing at present. 'Taking time to save time' – that's a load of cobblers. I do not have time to spit on occasions and neither do my guys, and this so-called TPM Pilot Team are now going to take a day per week to sit down and sort out the troubles of the world. How do they expect me to release people and run my shift?

The consultants that the company has brought in are costing us a fortune, but because they are from outside the management will listen to what they have to say. Us old timers have been telling them the same things for years, and they won't take any notice of us.

Well, I have just come back from a so-called coaching session on a one-to-one basis with a consultant and I honestly cannot see what we are paying for. This must be the biggest rip-off I have ever seen. They talk about teamwork but we do that already, otherwise we could not solve all the problems we come across. My shift have always worked together to problem solve, but the other shifts have a few strong personalities on them and they will never change.

This so-called Pilot Team have been meeting one day a week for the last six weeks and all I have seen is some form of fancy calculation called OEE – Overall Equipment Effectiveness. This is split up into Availability, Performance and Quality. They say this gives a more in-depth look at how your equipment is performing, but I cannot see anything wrong with the efficiency and quality defects that we report on each month at the moment. All this TPM is doing is creating more and more paperwork!

I have just come on shift on this Monday morning and we have already had a breakdown on the machine the TPM team are supposed to be working on. They have scheduled me to shut this machine off for a further four hours for them to do something called a Condition Appraisal. I have to hit my production targets and if they seriously think I am shutting down when I am already behind schedule, they have another think coming. Production comes first – that is what I am here for.

They have carried out that work on one of the other shifts and that cost us twelve hours' production downtime. For what?? No one has had any feedback as to what they've found, if they've found anything at all. They have put a few more pretty pictures up on their communications board, but they still have not achieved anything as far as I can see, although I must admit the machine is looking a lot cleaner.

I cannot believe what is happening in this company. That team has now had permission to shut down the machine for one week to carry out a 'Refurbishment Plan'. The number of times I have tried to improve this machine and I have not been allowed to because of 'production demands', is ridiculous, and here we have a group of 'shopfloor workers' being allowed to do so. They are undermining the authority of the Shift Team Leaders, as far as I am concerned.

That machine is now back in production and, apart from some initial teething problems, is running well. We have maintained these machines on many occasions in the past, and they run better for a while but then drift back to the condition they were in before they were worked on. They then become inconsistent and unreliable, and I can guarantee the same thing will happen with this machine.

The recognition phase

I have just come back from two weeks' holiday in Spain and am surprised to see that the TPM machine is still running well. I have also noticed a change in approach by the operators, who in some ways are now controlling the machine and not the other way round. They have changed their working routine and now carry out minor front line asset care tasks, which is all part and parcel of keeping their machines in good condition. They call this their Asset Care Routine. Conversations with these operators have changed, as they now have a great deal more knowledge of the process than they had in the past. The TPM process has now allowed the operators and maintainers to take on ownership of problems and allows them time to problem solve. The operators can now discuss more of the technical issues of the equipment, as they have a far greater understanding of the equipment. They have also become far more safety-conscious, as they now seem to understand how the kit actually works!

The change in relationships between skilled and unskilled is dramatic, and they are now discussing faults between themselves and coming up with answers without having to involve myself at all. The fitters and electricians have now trained the operators to carry out minor tasks that used to tie the skilled guys down on trivial jobs, so they have time to tackle the bigger problems on site. It has amazed me how people have taken this on board. In the past, if we as managers had asked the skilled personnel to train the operators, they would have gone on strike. As team members, they are doing it without question. The trouble with all this development of the shopfloor people is where does it leave me? All the time I have been a supervisor I have dictated to the shopfloor how they are expected to work on my shift. Now I have to back off and co-ordinate things rather than manage the people.

Collection of the data has been going on since the start of the pilot project and OEEs have been generated. I feel that this is the most difficult section of the nine-step process, as people cannot see how all the pieces of the jigsaw fit together at this early stage.

The buy-in phase

I have just undergone Team Leader training, which has given me a better understanding of how powerful a tool the OEE really is. The first OEE figures showed us that our machine was performing at an average of 35 per cent over a three-week period. We came up with a realistic figure as a target by taking the best availability, performance and quality for that three-week period, which came out at 52 per cent. This is classed as the 'best of best' and, if we can take control of our losses, we can achieve this figure consistently. When you actually put a cost on this, it equates to a saving of £150 K per annum, and if I find a fault I can justify downtime to eliminate that fault against loss of OEE per annum. This is the first time that I have actually been given a tool which allows me to go forward to upper management and be able to justify my spending and planned downtime on equipment.

The nine-step process is going well and the team have made their job a lot more user-friendly and also more efficient in the way they operate their equipment. This in turn gives me more time to complete my work without being disturbed and without people asking me to sort out their problems all the time.

We have also been trained in the art of CAN DO, which is the same as the Japanese 5S. We have had an initial clear-out of things that were not wanted or had been stored in our area by someone else. The neatness Step 2 is now being progressed and the place is looking a great deal better – it is surprising how much extra space we have created in the department. I have now been allocated an 'improvement zone' within my department which I am responsible and accountable for. I still run the whole department and have to ensure that we hit production targets as well as keeping the place clean, but I have one area as my TPM/improvement zone which includes CAN DO as well as attacking problems using the nine-step process.

The hardest problem is getting all shifts to buy into keeping each other's areas clean instead of dumping their rubbish into my area when they are due to be audited. We have had the same trouble with every initiative we have tried and if we can make it work across the five shifts, then we really have taken a massive step forward. At the moment, I feel that we are not going to move any further forward with this as there are a lot of people not wanting to change in this company. Luckily, I have never been that way inclined, or have I? The initial audits were set out with dates against when they were to be carried out. However, although we cleaned up the day before the audit, what we found was that the time taken to get the areas back to that standard was reduced dramatically. We have since changed to a system of auditing randomly, which means the department has to be kept at this level at all times.

Two years later – looking back

When I look back over the last two years and see the changes which have taken place within our organization, it has made a tremendous difference to the way I now do my job. During the initial introduction with the consultants supporting it, they stressed that it was not a quick-fix solution but a drip-feed change of attitudes and approach to problem solving. People like myself have changed our approach without actually realizing what was happening and, as I look at the upper management, the people who I felt would be the stumbling blocks when we started are no longer in the same positions as they were twelve months ago.

The main advantage I have found is that the TPM process is not just a machine-related system but becomes an all-encompassing approach. It looks at the door-to-door OEE, not just at an individual machine, but also at the value chain through the whole organization and involves everyone, from planning to finance to forklift truck driver.

The change in my role has allowed me to stand back, look at the bigger issues within our department, and not just be an ostrich/head in the sand type of Team Leader. The money we invested with consultants in the early days of TPM, with which I totally disagreed, has been paid back over and over. The problem we had, once we had an understanding of the process, was that TPM is not rocket science, it is an obsession with attention to detail, and we felt that we could do it all ourselves. You very quickly realize that you need someone to keep you on the right track, otherwise you get involved in side issues and forget what it is you are actually trying to achieve.

It is a complete change in culture and, carried out properly, it can be the best thing since sliced bread but, as the saying goes, 'You have to break a few eggs to make an omelette!'.

4
Techniques to deliver the TPM principles

The key significance of Seiici Nakajima's work in the evolution of TPM and the differences between the work ethic in Japan and that in the West have already been referred to in Chapter 1.
Nakajima established five pillars for the application of TPM:

1 Adopt improvement activities designed to increase the overall equipment effectiveness by attacking the six losses.
2 Improve existing planned and predictive maintenance systems (maintainer asset care).
3 Establish a level of self-maintenance and cleaning carried out by highly trained operators (operator asset care).
4 Increase the skills and motivation of operators and engineers by individual and group development (continuous skill development).
5 Initiate maintenance prevention techniques, including improved design procurement (early equipment management).

One of the main purposes of this book is to show linkages between techniques necessary to implement Nakajima's pillars by building on existing good practices. To reiterate the analogy: 'In a heart transplant operation, if you do not match the donor's heart to that of the recipient, you will get rejection'.
Nakajima's answer to the question 'What is TPM?' provides at least three basic aims:

- To double productivity, and reduce chronic losses to zero
- To create a bright, clean and pleasant factory
- To reinforce people (empower) and facilities and, through them, the organization itself

These aims are attractive to all, but the approach required will vary from one company to another. Experience has shown that tailoring TPM to the local plant-level organization and its people is the only way to achieve success. This process must be founded on the wide experience of applying TPM in different countries and in different industries, whilst at the same time recognizing local, plant-specific issues. An understanding of how TPM techniques link together is important to ensure that customization does not become cherry picking.
As explained in the previous chapter, the TPM improvement plan contains the techniques needed to apply the pillars or principles of TPM. This uses three cycles:

- measurement
- condition
- problem prevention

The present condition and future asset care requirements for the plant and equipment are first established and then developed through the measurement cycle, which sets the present and future levels of overall equipment effectiveness. Finally, the improvement cycle carries the process forward to the best of best and on to world class through a continuous improvement 'habit' (this concept is fully developed in Chapters 5 and 6).

Figure 4.1 shows the application of Nakajima's principles to the three-cycle improvement plan. Figure 4.2 shows the driving force behind Japanese TPM, and Figure 4.3 shows the approach pioneered by the Japanese Institute of Plant Maintenance.

	Measurement cycle	*Condition cycle*	*Problem prevention cycle*
1 Continuous improvement in OEE	✓	–	✓
2 Set up planned, preventive maintenance asset care	✓	✓	–
3 Establish operator asset care	✓	✓	✓
4 Continuous skill development	✓	✓	✓
5 Early equipment management	✓	–	✓

Figure 4.1 Relationship between five pillars (Nakajima) and three-cycle TPM improvement plan

Goal:	economic world domination via: – flexibility – right products – right time – right quality – right price
Trouble-free:	zero defects zero equipment failures zero accidents
Stockless:	no buffer stocks no WIP
All equals:	total waste elimination
TPM viewed as an essential pillar for equipment reliability and product repeatability through people and not the systems alone	

Figure 4.2 Essence of Japanese TPM

Total Productive Maintenance (TPM) combines the conventional practice of preventive maintenance with the concept of total employee involvement. The result is an innovative system for equipment maintenance that optimizes effectiveness, eliminates breakdowns and promotes autonomous operator maintenance through day-to-day activities.
Specifically, TPM aims at:

1. Establishing a company structure that will maximize production system effectiveness.
2. Putting together a practical shopfloor system to prevent losses before they occur, throughout the entire production system's life cycle, with a view to achieving zero accidents, zero defects and zero breakdowns.
3. Involving all departments, including production, development, sales and management.
4. Involving every single employee, from top management to front-line workers.
5. Achieving zero losses through small-group activities.

Figure 4.3 What is TPM? The JIPM definition

The scope for improving on the way we do things now can only be established by adopting the continuous improvement approach and by never accepting that what we are achieving today will be good enough for the future. A striking example of this comes from a visit by the authors some years ago to the press shop in a Toyota automobile plant in Japan, where it was observed that a 1500-tonne press die change took place in the astonishingly short time of $6^1/_2$ minutes. When this was commented on, the reply came: 'Yes, yes, we know, we need to reduce the time to 5 minutes.' At that time, a comparable change in a UK plant could take up to 4 hours. Straightforward die change is regularly achieved in a single minute!

Analogies and visual aids are essential components in the process of introducing TPM. One of these is the concept of healthy equipment, as already illustrated in Figure 3.12, which portrays the 'apple a day' for good health, the 'thermometer' to monitor well-being and the 'injection' to protect against disease. Routine asset care involving lubricating, cleaning, adjusting and inspecting ensures that the plant is protected against deterioration and that small warning signs are acted upon. Condition monitoring and prediction of impending trouble ensure that developing minor faults are never allowed to deteriorate to a breakdown or a reduced level of machine effectiveness. Finally, timely preventive maintenance safeguards against the losses which can come from breakdowns or unplanned stoppages. These messages are most effective when expressed in terms which hook into local and, hence, specific vision and values.

A key benefit of TPM, and an important strength of Japanese management, is the use of structured roles and responsibilities, which reduce both complexity and uncertainty. In reality, there is only one TPM. It is a package of integrated principles which are greater together than the sum of their individual parts. As companies improve, TPM has been adapted so that it continues to represent manufacturing's best practice. The original five principles remain at the core of the wider-reaching company-wide TPM discussed later in this chapter.

There are difficulties in implementing TPM in every country, including Japan. As this is the country with the most experience, the TPM implementation process is at its most mature in Japan. Naturally, this is an evolving situation as more non-Japanese companies achieve 'world-class' TPM applications. The three-cycle TPM improvement plan was developed to deal with the need to:

- progressively build management commitment and consensus based on results;
- build on existing good working practices;
- produce rapid results;
- get buy-in to new ideas across international boundaries.

Within the rigour of the three-cycle, nine-step process, it provides the flexibility to build on strengths and reduce weaknesses. In this way, it builds on the principles rather than diluting their undoubted synergy.

Let us now take a closer look at each of the five Nakajima TPM principles, together with the measurement, condition and problem prevention of the TPM improvement plan (Figure 4.1).

4.1 First principle: Continuous improvement in OEE

Figure 4.4 illustrates how the OEE links to the six losses. This demonstrates that central to the philosophy of TPM is the identification of reasons for the causes and effects of the six losses, such that their elimination is bound to lead to an improvement in the OEE. An example from the offshore oil industry shown in Figure 4.5 illustrates that poor asset care can lead to inadequate

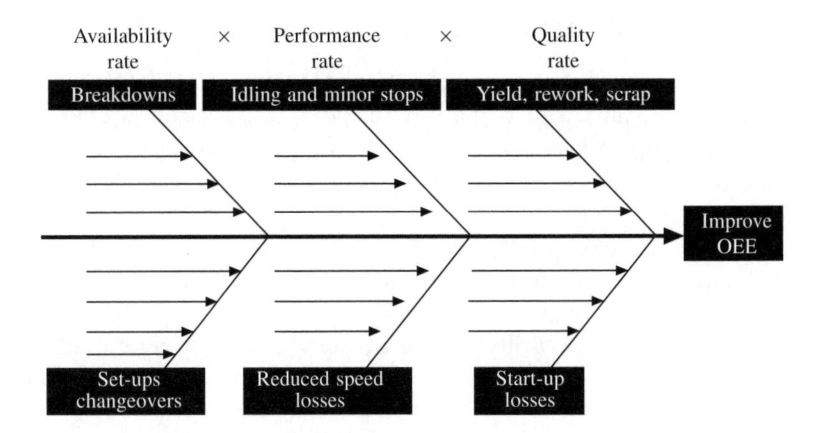

Figure 4.4 Factors in overall equipment effectiveness

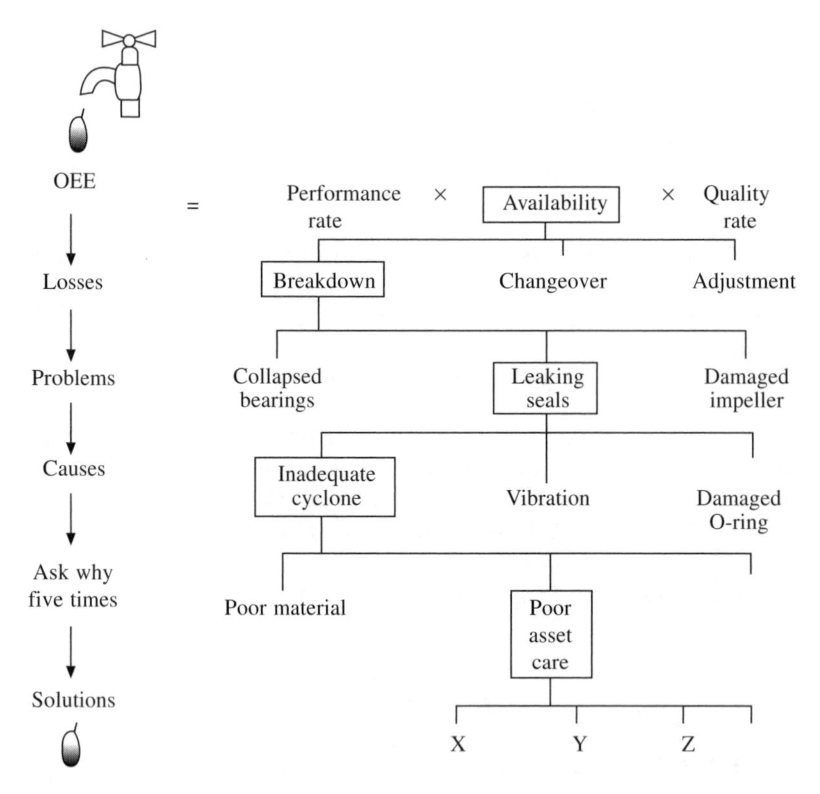

Figure 4.5 Problem-solving cascade

cyclone operation, which gives rise to a leaking seal problem, which results in a breakdown, which affects the availability part of the OEE measurement.

4.2 Second principle: improve planned, preventive maintenance

In the TPM improvement plan (Figure 3.19) this aspect is covered by steps 4 to 7 of the condition cycle. The key step is number 7, where the future asset care regime is determined by the TPM team, and is based on the principles illustrated in Figure 3.12.

The detailed approach to planned preventive maintenance will vary from plant to plant and from industry to industry, but an effective prevention-driven maintenance system is essential. However, unnecessarily intrusive maintenance should be avoided. If the equipment is running smoothly and no signs of defects or malfunctions are noticed in the daily routine of cleaning and inspection or through condition monitoring, then it is pointless to strip and rebuild the machine simply because it has been laid down as part of the maintenance plan. Highly sophisticated sensors and software are available to forestall failure in all types of rotating machinery, and they are non-intrusive. However, in TPM the best condition monitor is an operator who is in harmony with his equipment and who has a sense of ownership of that equipment. This leads on to the third principle.

4.3 Third principle: establish operator asset care (autonomous maintenance)

All three cycles of the TPM improvement plan (see Figure 4.1) involve the principle of autonomous maintenance, or operator asset care. In the Japanese TPM approach, there are seven steps of autonomous maintenance (see Figure 4.7). These are given below together with the necessary linkages to the three-cycle, nine-step TPM improvement plan.

Step 1 Initial cleaning

This starts with the 5 Ss mentioned in Chapter 1. The cleaning of machines and production plant gives operators an insight, which they never had before, into the condition of their machines. They can therefore use their eyes, ears, nose, mouth and hands to help their maintenance colleagues as an 'early warning system'. By working together as a team they can ensure effective asset care and release maintenance people for tasks requiring a higher level of training and skill. The full implications of the cleaning regime cannot be over-emphasized because ultimately it leads to the reform of the whole production process. To understand this clearly it pays us to look again at the Japanese 5 Ss and the 'localized' CAN DO approach, and the way in which their application leads to fundamental changes in the workplace.

The Japanese 5 Ss emphasize the concept of keeping things in the workplace under control.

Seiri (organization)

This is the practice of dividing needed and unneeded items at the job site and quickly removing the unneeded ones. It also means integrating material flow with the best known operational methods.

To better understand the meaning of unneeded items, these can be divided into three different categories:

- Defective products
- Not useful items
- Not urgent objects, right now

There are six recommended categories in *seiri* with their own targets for improvement:

- Stock, inventory
- Tools, jigs
- Dies

- Containers, pallets
- Conveyors, trucks, forklifts
- Space

Seiton (orderliness)

This means orderly storage, putting things in the right place. Those things can then easily be found, taken out and used again when they are needed. It doesn't simply mean lining things up neatly; it means there is a place for everything, and everything should be in its place! The locations of equipment, tooling and materials are clearly defined, displayed and maintained.

Seiso (cleaning)

This refers to cleaning the workplace regularly, to make work easier and to maintain a safe workplace.

Seiketsu (cleanliness)

This means being aware of the need for maintaining a clean workplace, not just through cleaning programmes but through ensuring that spillage of liquids and dropping of materials, packaging, etc. is avoided.

Shitsuke (discipline)

This means to formalize and practise the above items continuously each day as you work, to have the discipline to always work to these principles.

In WCS International we have developed an eleven-step plant-wide clear and clean exercise for our clients as a start point to put the philosophy of 5S or CAN DO into practice. This is often implemented *shortly after initial TPM pilot equipment projects have been launched*, in order to get everyone involved at an early stage. It is not used as a forerunner of TPM, as is the usual case with the Japanese approach. The Japanese seem quite prepared to spend six to twelve months cleaning up a plant. In the Western world we do not quite have the same level of patience, and we need to experience early live equipment examples called *pilots* in order to illustrate, prove and believe in the TPM process.

The initial plant clear and clean process is described as follows:

Clear out

1. Zone the plant into clear geographical areas with clear management responsibility. (See the plant plan for your shift's responsibility area.)
2. Carry out a first-cut physical run for items that can be immediately thrown away *today* because it is *obvious* they are not needed.
3. Carry out a second red-tag/red-label/red-sticker run, which needs to be more structured and thoughtful.
4. It is obvious that if you are to get rid of a great many items, you will need a great many waste disposal containers (say six strategically placed skips). Some items will be wanted but are in the wrong place: 'There must be a place for everything, and everything must be in its right place.'

5 For things to be in the right place we need to paint clear gangways and clear markings on the floor for anything mobile (i.e. sillages, raw material, work in progress, etc.). Correct racking, shadow boards, labelling and other visual storage aids will form an important part of this stage.
6 Keep the workplace organization under a permanent microscope.

Clean up

7 Do the obvious sweeping and vacuuming of the work area.
8 Inspect and clean every square centimetre of the equipment. Remember: *every square centimetre.*
9 Identify the points of accelerated deterioration. Where are the leakages and spillages occurring, and *why*? Ask 'why?' five times.
10 Get to the *root causes* of dust, dirt and scattering and *eliminate* those reasons. We will achieve a dust-free plant if – and only if – we achieve this step. All the previous nine steps are useless unless step 10 is achieved.
11 Revisit steps 1 to 10 and continuously improve.

Step 2 Countermeasures at the source of the problems

Cleaning, checking, oiling, tightening and alignment of equipment on a daily basis enable operators to detect abnormalities as soon as they appear. From then on, operators learn to detect problems and to understand the principles and procedures of equipment improvement. To set this in perspective, we can list some examples of situations where the operators have *not* been trained to be equipment-conscious:

- dirty or neglected equipment
- disconnected hoses
- missing nuts and bolts, producing visible instability
- steam leaks and air leaks
- air filter drains in need of cleaning
- jammed valves
- hydraulic fluid and lubricating oil leaks
- measuring instruments too dirty to read
- abnormal noises in pumps and compressors

These are glaring examples of a failure to maintain the most basic equipment conditions, but we are deluding ourselves if we believe such situations never arise – they do! Even brand new equipment, if neglected, will rapidly deteriorate (i.e. after just a few days) and its performance and output will drop as a consequence.

Use of visual management techniques

When the equipment has been cleaned and the weaknesses have been found and corrected, the next phase of the TPM process is to draw attention to the

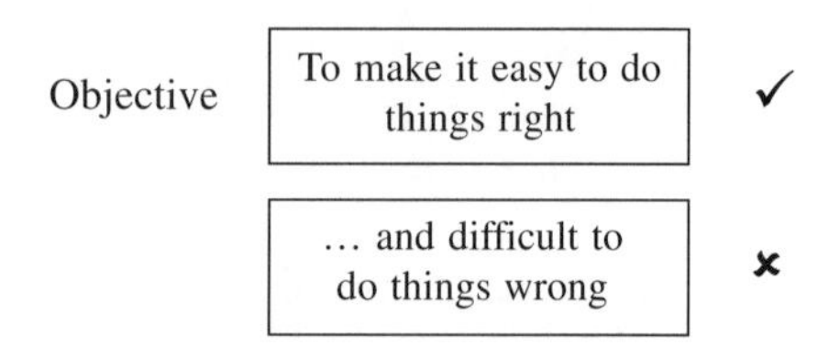

Figure 4.6 Improvements

right way of doing things by clear visual aids. This is *error-proofing*: to make it easy to do things right and difficult to do things wrong (Figure 4.6). Some examples of visual marking to encourage ease of inspection, discipline, order and tidiness are as follows:

- Where sight glasses are used, make sure that they are clean and that the high and low points are boldly marked and colour coded so that they can be seen at a glance.
- Mark gauges green for 'go' and red for 'no go'.
- Use small windmills to indicate extraction fans and motors working.
- Indicate the correct level on oil bottles as a maximum and a minimum. An elastic band on the bottle will show the level at the previous check, to give the *rate* of use.
- Use line indicators on bolts and nuts to show position relative to their fixture base.
- Provide inspection windows for critical moving parts.
- Colour tag clearly those valves which are open and those which are closed.
- Highlight critical areas which must be kept scrupulously clean.
- Identify covers which are removable by colour coding them.
- When there is an agreed inspection routine, number in sequence those points which require attention.
- Prepare quality colour photographs of equipment standards and ensure that these are readily accessible to operators.
- Make up shadow boards for tools and spares so that the correct location of every item is immediately apparent.
- Indicate the correct operation of machines by instructions and labels which are visible *on the machine*, kept clean and accessible.
- Display charts and graphs adjacent to the equipment to show standards and to indicate progress towards objectives.

Having completed the first two steps towards autonomous maintenance, operators will have learned to detect problems and to understand the principles and procedures of equipment improvement. They can now take the next steps.

Step 3 Cleaning and lubrication standards

Much will have been learned from the initial cleaning, orderliness and discipline procedures, and it will now be possible to set standards for the ongoing care of plant and machines. This will lead logically towards the next step.

Step 4 General inspection

Helped by other members of the team, operators can be guided towards the point where they can carry out general inspection themselves. They will then have reached the stage where they know the function and structure of the equipment and have acquired the self-confidence to make a much more significant contribution towards the goal of more reliable machines and better products.

Step 5 Autonomous inspection

As the term implies, operators can now carry out self-directed inspection routines and repairs/servicing as required.

Step 6 Organization and tidiness

Initial cleaning and the application of the 5S/CAN DO philosophy will by now have worked through and started to have major effects. In parallel with this, operators will have reached the stage where they can take responsibility for performing autonomous inspection – always within the limits of their skills, experience and training and always backed, where necessary, by their maintenance colleagues. They will have developed an understanding of the relationship between equipment accuracy and product quality. This leads to the final step.

Step 7 Full autonomous maintenance

At this stage, operators will be equipped to maintain their own equipment. This will include cleaning, checking, lubricating, attending to fixtures and precision checking on a daily basis. They are now equipped to apply their newly developed skills and knowledge to the vital task of continuous improvement.

The key point of emphasis in developing these asset care routines is *empowerment*. The operators' and maintainers' own ideas are encouraged and adopted on the basis that 'If it is my idea and it is embodied in the way in which we operate and look after our equipment, then I will stick with it!' On the other hand, 'If it is imposed from above, then I might tick a few check boxes, but I won't actually do anything!' The progress from cleaning to full autonomous maintenance is illustrated in Figure 4.7.

The condition cycle of the TPM improvement plan (Figure 3.19) moves through the following steps:

- Criticality assessment
- Condition appraisal
- Refurbishment programme
- Future asset care

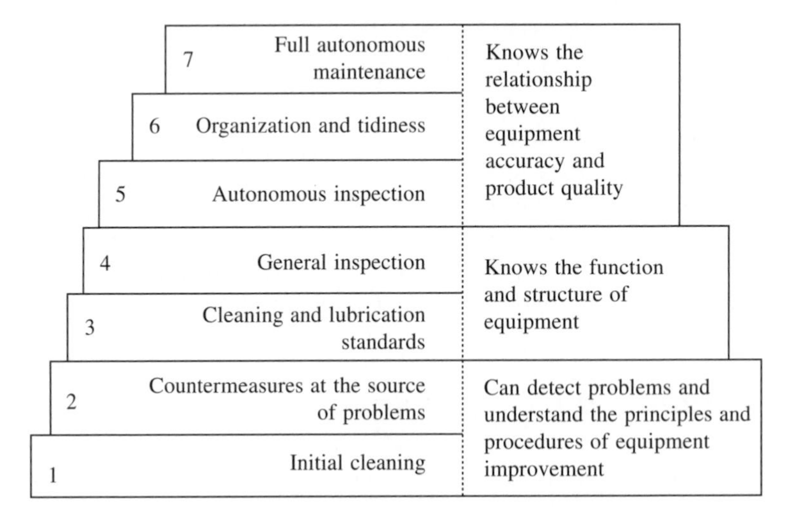

Figure 4.7 Seven steps for developing autonomous maintenance

These cover the seven steps of autonomous maintenance, but provide the structure and discipline to link the process to progress the other principles incorporated within the measurement and problem prevention cycles.

One of the key points that the reader should now appreciate is that no rocket science is involved in the concept of TPM: it is basically sound common sense.

The reality of implementation is that both autonomous maintenance and planned maintenance must be implemented in parallel. This is why WCS have developed the concept of asset care to formally integrate production and maintenance stepwise improvement. Unless the actions of the other pillars support their efforts, progress will be slow.

4.4 Fourth principle: continuous skill development

The whole emphasis of TPM, and hence the three-cycle, nine-step improvement plan, is geared to taking its participants – whether the Chief Executive or the operators and maintainers – on a journey from innocence to excellence (Figure 4.8).

Training is about learning and understanding. The best and only way to *retain* our learning and understanding is through *experience*. Learning is best retained by a series of well-structured and relevant single-point lessons. There is only one way to learn TPM, and that is to actually go and *do it*! However, preparation and awareness are also all-important before putting it into practice. So we have to proceed with thought, planning, care and 100 per cent attention to detail.

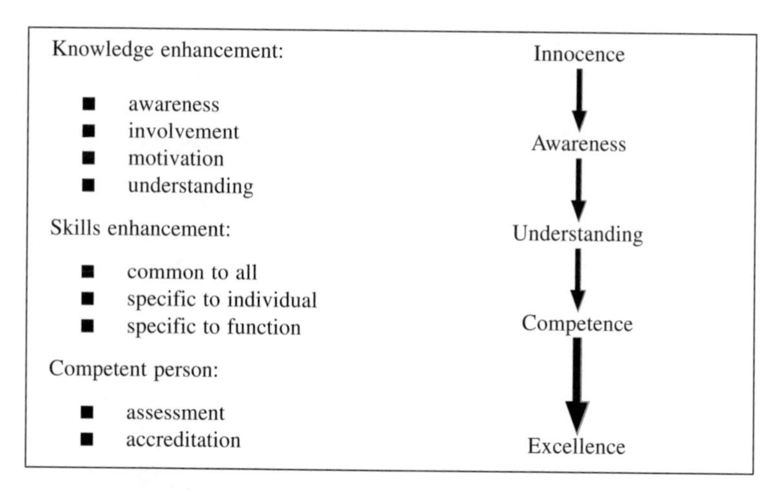

Figure 4.8 TPM awareness and training: innocence to excellence

We have said already, and it is worth repeating: TPM is a philosophy or a concept which uses hard and proven enabling tools. Moreover, TPM can be measured. It is total quality with teeth! It is also a powerful involvement process which has to be communicated through awareness and training. The benefits of training are not always immediately apparent, but if you believe training is expensive, try ignorance: that can be really expensive!

Remember also that communication is a two-way continuous process, and so is TPM. As a TPM facilitator at Vauxhall Motors, Ellesmere Port, said at the completion of his TPM pilot team presentation: 'TPM stands for "today people matter", and as a result you will get "totally pampered machines" '.

In setting up your TPM programme, you should give clear thought and definition to the following:

1 Establish purpose of training.
2 Establish training objectives.
3 Agree method of delivery.
4 Set up training framework and modules.
5 Design training and awareness programme.

The total training programme can result only from detailed study of a particular company or plant, and the programme must be tailored to the needs of all the people involved and applied to all equipment they are working with. Chapter 7 gives full consideration to designing your own tailored TPM training programme.

4.5 Fifth principle: early equipment management

Equipment management concerns the complete equipment life cycle from concept to disposal. The principle of Early Equipment Management recognizes the importance of the early stages in the reduction of life cycle costs.

This principle is implemented using three TPM for Design (TPM (D)) techniques (see Table 4.1), each of which is directly linked to the improvement plan outputs.

Unless good equipment management skills are nurtured, the designers will not understand how to use shopfloor information, no matter how good it is. Furthermore, if the designers are not skilled enough to recognize operational weaknesses, they will not be able to create effective designs. Most designers have little work experience in equipment operation and maintenance, so they do not think in terms of operability and maintainability. However, they can overcome these weaknesses and build equipment design skills by:

- visiting the factory floor and hearing what the equipment operators and maintenance staff have to say;
- studying equipment that has been improved as a result of autonomous maintenance or quality maintenance activities and listening to project result announcements made by TPM circles;
- getting hands-on experience in cleaning, lubricating and inspecting equipment;
- conducting several P–M analyses based on checklists.

Designers should have their knowledge and skills evaluated in order to identify remaining weaknesses, facilitate self-improvement, and acquire on-the-job training in more advanced skills.

There is an extraordinarily powerful commercial advantage to a company when this vital pillar and principle of TPM can be mobilized and used to maximum effect. Designers, engineers, technologists, procurement, finance, operations and maintenance will then work as essential partners in the drive to improve the company's overall equipment effectiveness by eliminating many of the reasons for poor maintainability, operability and reliability at source (i.e. at the equipment design, engineering and procurement stage).

Table 4.1 TPM (D) links with 9 step TPM improvement plan

Component	*Purpose*	*Link with improvement plan*
Design process milestones and organization	To co-ordinate the parallel activities of commercial, engineering and operations functions	Measurement cycle Assessment of loss Prioritization/targets
Evolution of the design knowledge base	To co-ordinate transfer of lessons learnt and adoption of best practice routines	Step 8 Best practice evolution
Objective testing	To select equipment options based on evidence of suitability	Results from Step 9 problem prevention activities

The infrastructure of core TPM teams supported by key contacts (see Figure 3.13) reflects the importance of TPM (D) both in retrofit on existing assets and for the next generation of equipment. Table 4.1 shows that this vital pillar is also focused through the measurement and problem prevention cycles of the nine-step TPM improvement plan. Chapter 9 is devoted to the subject of TPM for equipment designers, specifiers and planners.

4.6 Company-wide TPM

As equipment losses are brought more under control, management attention can be directed towards other issues in the value stream/supply chain.

The progression of classic TPM into company-wide TPM recognized the need to define more closely the changing roles of maintainers and managers as they evolve from looking inward to becoming customer-focused.

The importance of safety is also given recognition as well as TPM in the office environment. The number of pillar champions under company-wide TPM has increased from five to eight.

The classic five principles are a good starting point for implementing top-down roles. These correspond closely to key management roles (Table 4.2).

Management's role in TPM is to create an environment which pulls through continuous improvement bottom-up rather than pushing initiatives top-down. A good way of explaining the five pillar champion roles is by linking the pillar champion to the five questions set out in Chapter 3.

Champion	*Question*
OEE	Why don't we know the consequences of failure (both obvious and hidden)?
OAC	Why does this part of the process not work as it is meant to?
MAC	Why can't we improve the reliability?
CSD	Why don't we have the skills to set optimal conditions?
EEM	Why can't we maintain and improve our technology to maintain optimal conditions for longer?

Table 4.2 Pillar champion/management roles

Pillar champion	*Management role*
Continuous improvement in OEE	General Manager
Operator asset care	Manufacturing Manager
Maintainer asset care	Engineering Manager
Continuous skill development	HR/Personnel Manager
Early equipment management	Research and Development/Design Manager

Promoting the environment where those questions are asked and answers sought throughout the organization is an essential leadership task which needs to be allocated sufficient of the following in order to succeed:

- Priority
- Pace
- Resources

Experience has shown that the programme should focus initially on the classic five pillars. The evolution from five pillars to eight can take place as the first milestone (everyone involved) is passed (Figure 4.9).

The additional pillars are described below. These build on the change from *reactive* to *proactive* management potential released by achieving zero breakdowns. They support problem finding and the progressive reduction of chronic losses.

Focused improvement

Top-down driven improvement directed at issues identified bottom-up. Key contacts working with and on behalf of teams to support resolution of support problems/improvements.

Quality maintenance

The development route for maintainers analysing minor quality defects to identify and pursue optimum conditions and chronic loss reduction. Provides an essential bridge between today's shop-floor learning and the early equipment knowledge base.

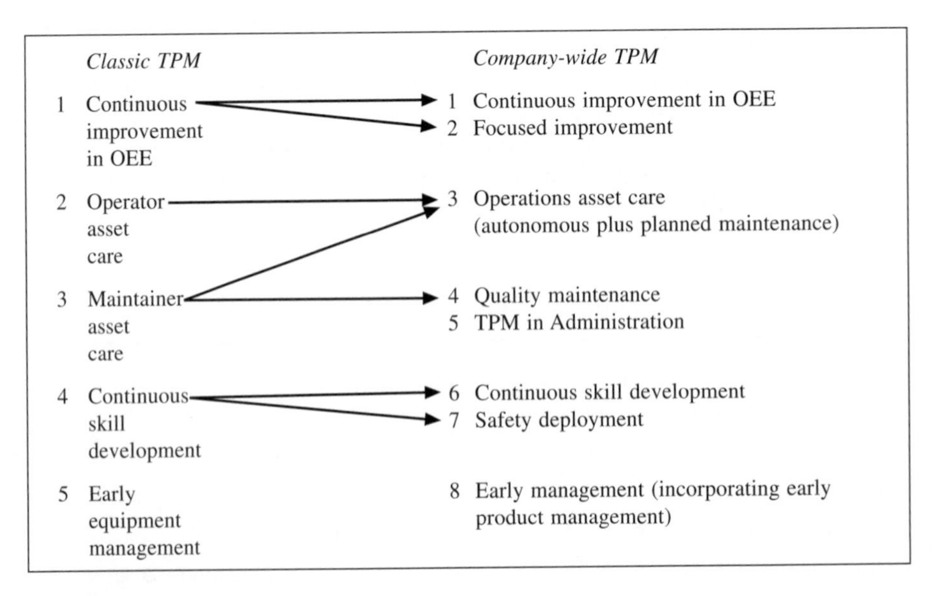

Figure 4.9 Evolution of pillars

TPM in administration

The application of loss identification and reduction techniques to systems activities and tasks. A similar approach to the way TPM tackles plant, lines and equipment.

Safety deployment

Focus on risk reduction by making the easy way the right way. Incorporates behavioural safety and the use of visual techniques and competence-based training to reduce the amount of written information needed to ensure environmental and personal safety.

5
The TPM improvement plan

TPM is about maximizing the overall effectiveness of equipment through the people who operate and maintain that equipment. In order to provide the essential link between equipment and people it is essential to identify a clear set of phases and steps which together make up the TPM improvement plan. As outlined in Chapter 3, there are three phases to the plan:

- The *measurement cycle*, which assesses the present effectiveness of the equipment and provides a baseline for the measurement of future improvement.
- The *condition cycle*, which establishes the present condition of the equipment and identifies the areas for improvement and future asset care.
- The *problem prevention* cycle, which moves equipment effectiveness forward along the road to world-class performance.

For convenience, Figure 5.1 repeats Figure 3.6.

Throughout the TPM improvement plan five themes prevail, as discussed in Chapter 3.

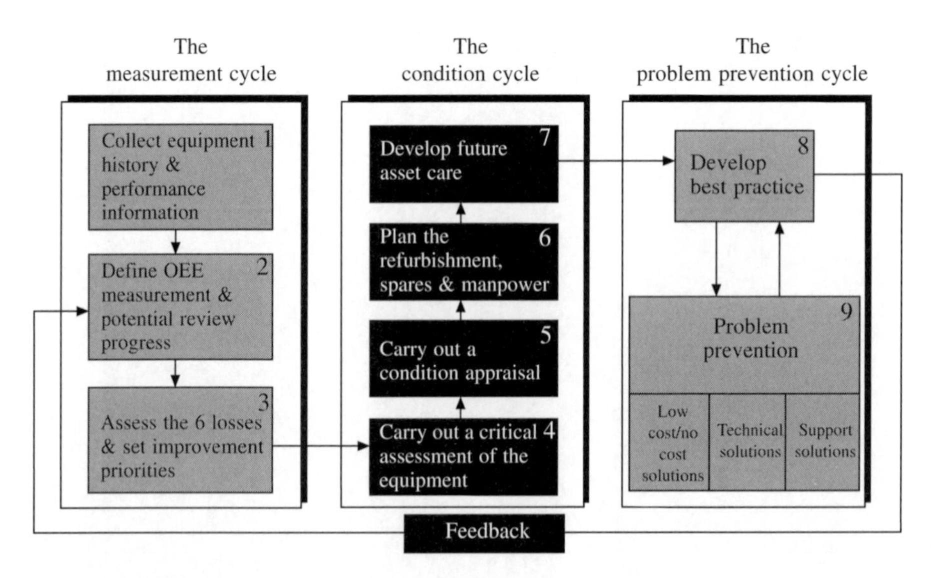

Figure 5.1 TPM improvement plan: three phases and nine steps

1 Restore before improve.
2 Pursue ideal conditions.
3 Eliminate minor defects.
4 Prevent causes of dust, dirt and scattering.
5 Ask why five times:

- Why don't we know the true consequences of failure (both obvious and hidden)?
- Why does this part of the process not work as it is meant to?
- Why can't we improve the reliability?
- Why can't we set the optimal conditions for the process?
- Why can't we maintain those optimal conditions?

5.1 Measurement cycle

Step 1 Equipment history and performance record

This is the essential prerequisite to the OEE calculation (Step 2) because it records the recent effectiveness of an equipment item. This forms a basis for the problem prevention cycle (see later).
Included in this record are:

- data on equipment availability, performance and quality to enable OEE to be calculated;
- records of problems and breakdowns as a basis for problem solving and as evidence of improvements resulting from refurbishment and ongoing asset care;
- measurements and records of pressure, noise, vibration and temperature to show up any adverse trends. Under this heading will come data collection and data analysis from condition monitoring equipment.

All of this information will have a direct bearing on the ongoing asset care and improvement programme.

A typical equipment history record is shown in Figures 5.2 and 5.3, and records of this type will form the basis of the OEE calculation described below.

Step 2 Overall equipment effectiveness

The OEE formula is at the heart of the TPM process. It is soundly based on measurable quantities and enables progress to be quantified as the organization embraces TPM with all its implications. The formula enables calculation of two parameters:

- *Actual effectiveness* of the equipment, taking into consideration its availability, its performance rate when running and the quality rate of the product produced. All of these are measured over a period.

Total Productive Maintenance Programme													
R/H Front Door Line CO_2 MIG Welding Cell													
Material Usage Chart													
Material	*Week No*	*Week No*	*Week No*	*Week No*	*Week No*	*Week No*	*Week No*	*Week No*	*Week No*	*Week No*	*Week No*	*Week No*	*Week No*
Welding wire													
Welding tip													
Shroud													
Anti-spatter spray													
Distilled water													

Figure 5.2 Material usage chart

Equipment	*CO_2 MIG welding cell*				*VIII Ideal cycle time: 0.5 min/piece*		
	I	II	III	IV	V	VI	VII
Date/Shift	*Total working time* (min)	*Planned downtime* (min)	*Total available time* (min)	*Actual downtime* (min)	*Output* (no.)	*Defects (rework)* (no.)	*Defects (scrap)* (no.)
28.9.99 D/S	240	20	220*	5	243	0*	0*
28.9.99 N/S	240	20	220	0*	380	0	0
29.9.99 D/S	240	20	220	10	328	0	0
29.9.99 N/S	240	20	220	5	200	0	0
30.9.99 D/S	240	20	220	0	326	0	0
30.9.99 N/S	240	20	220	10	345	0	0
1.10.99 D/S	240	20	220	10	103	0	0
1.10.99 N/S	240	20	220	5	386*	0	0
2.10.99 D/S	240	20	220	5	187	0	0
Total	2160	180	1980	50	2498	0	0

* Best scores for use in calculation of best of best

Figure 5.3 Equipment history record for OEE calculation

- *Potential improvement* The first improvement objective is to obtain consistently, through standardization and stabilization, the best of the best in each of the three categories: availability, performance and quality. Beyond this point there must be continuous improvement towards world-class levels.

The OEE formula is as follows:

Overall equipment effectiveness	=	availability of the asset	×	performance rate when running	×	quality rate of product produced

The three factors in the OEE calculation are all affected to various degrees by the six big losses which were outlined in Chapter 3 and are shown in Figure 5.4. It is only by a single-minded and sustained attack on these losses that the TPM process can become effective – a change that will be demonstrated by improvements in the OEE.

- *Availability* will be affected by breakdown losses and by set-up (or changeover) and adjustment losses. (A breakdown requires the presence of a maintenance engineer to correct it.)
- *Performance* will be affected by idling and minor stoppage losses and by reduced speed losses. (A minor stoppage can be corrected by the operator and is usually of less than 10 minutes' duration.)
- *Quality* will be affected by defect and rework losses and by start-up losses.

In detail, the calculation of the three factors whose product determines the OEE is as follows:

$$\text{Availability} = \frac{\text{total available time} - \text{actual downtime}}{\text{total available time}} \times 100\%$$

Total available time is total attendance time less downtime allowances (i.e. tea breaks, lunch breaks) and planned downtime. Unplanned downtime is caused by breakdowns and set-ups and changeovers.

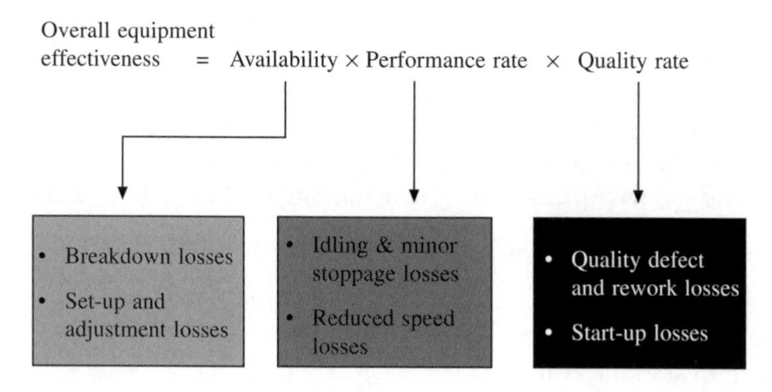

Figure 5.4 OEE and the six losses

Performance = operating speed rate × operating rate

$$= \frac{\text{Ideal cycle time}}{\text{Actual cycle time}} \times \frac{\text{actual cycle time} \times \text{output}}{\text{operating time}} \times 100\%$$

Ideal cycle time is the cycle time the machine was designed to achieve at 100 per cent. Output is output including defects. Operating time is total available time minus unplanned stoppages (i.e. available time).

$$\text{Quality} = \frac{\text{total output} - \text{number of defects}}{\text{total output}} \times 100\%$$

OEE calculation for welding cell

Calculation of OEE can best be demonstrated by using the values in Figure 5.3. The roman numerals refer to the columns in the figure.

Average OEE calculation

$$\text{Availability} = \frac{\text{III} - \text{IV}}{\text{III}} = \frac{1980 - 50}{1980} \times 100 = 97.5\%$$

$$\text{Performance} = \frac{\text{V} \times \text{VIII}}{\text{III} - \text{IV}} = \frac{2498 \times 0.5}{1980 - 50} \times 100 = 64.7\%$$

$$\text{Quality} = \frac{\text{V} - \text{VI} - \text{VII}}{\text{V}} = \frac{2498 - 0 - 0}{2498} \times 100 = 100\%$$

Average OEE = 0.975 × 0.647 × 1.000 × 100 = 63.1%

Best of best (target) OEE calculation

The best of best calculation uses the best scores in the period from each column. This gives us a theoretical achievable performance if all of these best scores were consistently achieved. It is our first target for improvement.

Best of best OEE = 1.000 × 0.877 × 1.000 × 100 = 87.7%

Question What is stopping us achieving the best of best consistently?
Answer We are not in control of the six big losses!

The best of best calculation generates a high confidence level, as each value used of the three elements (availability, performance, quality) was achieved at least once during the measurement period. Therefore, if control of the six big losses can be achieved, our OEE will be at least the best of best level.

We can now start putting a value to achieving the best of best performance.

TPM potential savings for achieving best of best

Cycle time A = 30s
Number of men B = 2
Allowance in standard hours
(lunch breaks, technical allowance, etc.) C = 11%

Credit hours generated per piece

$$X = \frac{(A \times B) + C}{3600s} = \frac{(30 \times 2) + 11\%}{3600s} = 0.0185$$

Variable cost per credit hour Y = £27.50
Direct labour cost per price X × Y = £0.5106
Current OEE D = 63.1%
Number of pieces produced E = 2498
Best of best OEE F = 87.7%

Number of pieces produced at OEE = 87.7% $G = \frac{F}{D} \times E = 3472$

Difference in pieces produced G – E = 974
Potential weekly savings = £0.5106 × 974 = £497
Potential annual savings (45 working weeks) = £22 365

An alternative to increasing the output potential of 974 pieces per week at best of best is to achieve the same output of 2498 pieces in less time:

Loading time (total available time) was 1980 minutes (33 hours) to produce 2498 pieces at OEE of 63.1 per cent.

Loading time to produce 2498 pieces at best of best OEE of 87.7 per cent would be:

$$\frac{63.1}{87.7} \times 33 = 23.74 \text{ hours} = 1425 \text{ minutes}$$

$$\text{Time saving} = 1980 - 1425 = 555 \text{ minutes} = 9.25 \text{ hours}$$

Simple OEE calculation

If the foregoing 'live' example seemed a little complicated, let us take the following very simple example to illustrate the principles.

Data

- Loading time = 100 hours, unplanned downtime = 10 hours
- During remaining run time of 90 hours, output planned to be 1000 units. We actually processed 900 units
- Of these 900 units processed, only 800 were good or right first time
- What is our OEE score?

Interpretation

Availability: actual 90 hours out of expected 100 hours
Performance: actual 900 units out of expected 1000 units in the 90 hours
Quality: actual 800 units out of expected 900 units

Calculations

Planned run time a = 100 hours
Actual run time b = 90 hours

(owing to breakdowns, set-ups)

Expected output in actual run time c = 1000 units in the 90 hours
Actual output d = 900 units

(owing to reduced speed, minor stoppages)

Expected quality output e = 900 units
Actual quality output f = 800 units

(owing to scrap, rework, start-up losses)

$$\text{OEE} = \frac{b}{a} \times \frac{d}{c} \times \frac{f}{e} = \frac{90}{100} \times \frac{900}{1000} \times \frac{800}{900} = 72\%$$

OEE calculation for an automated press line

Working pattern

- Three shifts of 8 hours, 5 days per week
- Tea breaks of 24 minutes per shift

Data for week

- 15 breakdown events totalling 43 hours
- die changes averaging 4 hours each per set-up and changeover
- 15 500 units produced, plus 80 units scrapped, plus 150 units requiring rework
- Allowed time as planned and issued by production control for the five jobs was 52 hours, including 15 hours for set-up and changeover

OEE for week

Loading time = attendance – tea breaks = 120 – 6 = 114 hours

Downtime = breakdowns + set-ups and changeovers

= 43 + 20 = 63 hours

$$\text{Availability} = \frac{114 - 63}{114} = 44.7\%$$

Actual press running time (uptime) = 120 – 6 – 43 – 20 = 51 hours

Allowed press running time = 52 – 15 = 37 hours

$$\text{Performance rate} = \frac{37}{51} = 72.5\%$$

Product input (units) = 15 500 + 80 + 150 = 15 730

Quality (first time) product output (units) = 15 500

$$\text{Quality rate} = \frac{15\,500}{15\,730} = 98.5\%$$

$$\text{OEE} = 0.447 \times 0.725 \times 0.985 = 31.9\%$$

Data for four-week period
Over a recent four-week period the following OEE results were obtained:

Week	*OEE (%)*	=	*Availability (%)*	×	*Performance rate (%)*	×	*Quality rate (%)*
1	44.6	=	65.0	×	70.0	×	98.0
2	43.8	=	58.0	×	77.0	×	98.0
3	36.7	=	47.0	×	80.0	×	97.5
4	31.9	=	44.7	×	72.5	×	98.5
Average	39.4	=	53.7	×	74.9	×	98.0

Best of best OEE and potential benefit
The best of best OEE can now be calculated. In addition, if the hourly rate of added value is taken to be £100, the annual benefit (45-week year) of moving from the current average OEE of 39.4 per cent to the best of best can be found.

$$\text{Best of best OEE} = \text{availability} \times \text{performance} \times \text{quality}$$

$$= 65.0 \times 80.0 \times 98.5 = 51.2\%$$

Potential loading hours per year = 114 × 45 = 5130

At 39.4% OEE, value added per year = 0.394 × 5130 × £100 = £202 122
At 51.2% OEE, value added per year = 0.512 × 5130 × £100 = £262 656

Therefore, a benefit of £60 534 is possible by consistently achieving best of best through tackling the six losses using the nine-step TPM improvement plan.

Step 3 Assessment of the six big losses

The importance of understanding and tackling the six big losses cannot be over-emphasized! They were listed in Chapter 3 and illustrated by the iceberg analogy in Figure 3.14, repeated here as Figure 5.5. The six losses are as follows:

- Breakdowns
- Set-up and adjustment
- Idling and minor stoppages
- Running at reduced speed
- Quality defect and rework
- Start-up losses

These are elaborated in Figures 5.7–5.12 in terms of the relationship of these losses to the OEE.

Figure 5.6 shows the losses as a fishbone cause and effect diagram. This formula is used by the TPM core team as a brainstorming tool to list all possible causes and reasons for each of the six loss categories.

We will develop a detailed definition in later chapters regarding the four levels of control referred to under each of the six losses in Figures 5.7–5.12. However, in order to give an early indication a definition is as follows:

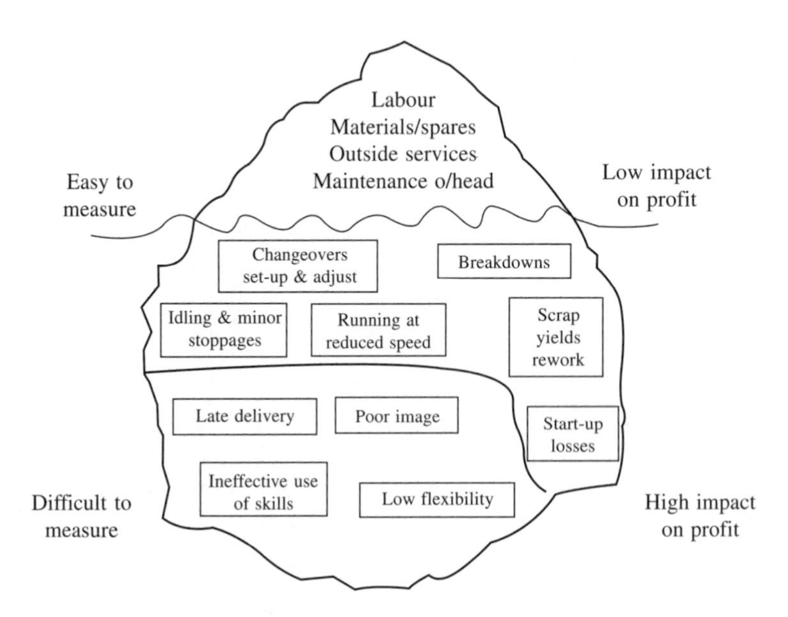

Figure 5.5 True cost of manufacturing: seven-eighths hidden

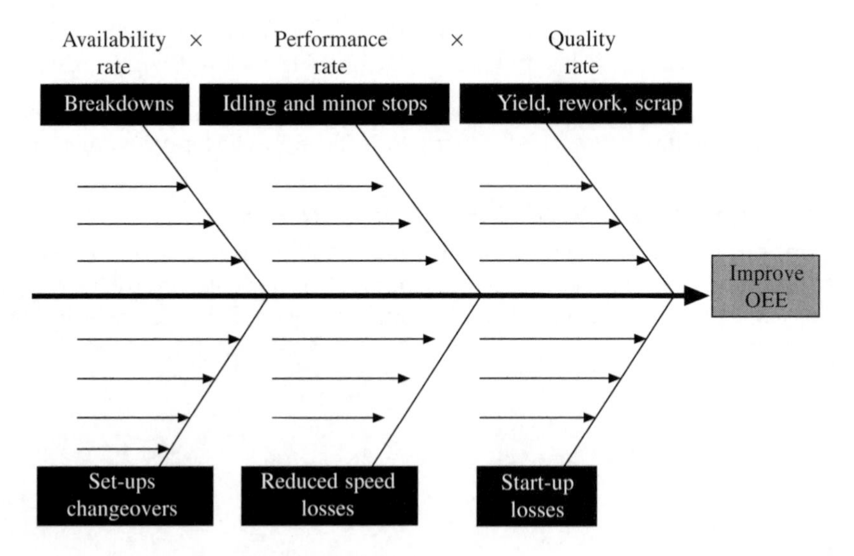

Figure 5.6 Factors in overall equipment effectiveness

- *Level 1* Milestone 1 after pilot/roll-out activity: 12–18 months
- *Level 2* Refine best practice and standardize: 6–12 months later (P–M prize level)
- *Level 3* Build capability: 12–18 months later
- *Level 4* Strive for zero: 3–5 years from roll-out launch

Level 1	*Level 2*
1 Combination of sporadic and chronic breakdowns 2 Significant breakdown losses 3 BM > PM 4 No operator asset care 5 Unstable lifespans 6 Equipment weaknesses not recognized	1 Chronic breakdowns 2 Breakdown losses still significant 3 PM = BM 4 Operator asset care implemented 5 Parts lifespans estimated 6 Equipment weaknesses well acknowledged 7 Maintainability improvement applied on above points
Level 3	*Level 4*
1 Time-based maintenance 2 PM>BM 3 Breakdown losses less than 1% 4 Autonomous maintenance activities well established 5 Parts lifespans lengthened 6 Designers and engineers involved in higher-level improvements	1 Condition-based maintenance established 2 PM only 3 Breakdown losses from 0.1% to zero 4 Autonomous maintenance activities stable and refined 5 Parts lifespans predicted 6 Reliable and maintainable design developed

BM Breakdown Maintenance PM Predictive Maintenance

Figure 5.7 OEE assessment: breakdown losses

Level 1	*Level 2*
1 No control: minimum involvement by operators 2 Work procedures disorganized: set-up and adjustment time varies widely and randomly	1 Work procedures organized, e.g. internal and external set-up distinguished 2 Set-up and adjustment time still unstable 3 Problems to be improved are identified
Level 3	*Level 4*
1 Internal set-up operations moved into external set-up time 2 Adjustment mechanisms identified and well understood 3 Error-proofing introduced	1 Set-up time less than 10 minutes 2 Immediate product changeover by eliminating adjustment

Figure 5.8 OEE assessment: set-up and adjustment losses

The improvement cycle in TPM starts from an appreciation of what the six big losses are and proceeds through problem solving to the establishment of best practice routines. Eliminating the root causes of the six losses is tackled in Step 9 of the TPM improvement plan.

Finally, Figure 5.13 shows a summary of the loss categories with improvement strategy examples.

Level 1	*Level 2*
1 Losses from minor stoppages unrecognized and unrecorded 2 Unstable operating conditions due to fluctuation in frequency and location of losses	1 Minor stoppage losses analysed quantitatively by: frequency and location of occurrence; volume lost 2 Losses categorized and analysed; preventive measures taken on trial and error basis
Level 3	*Level 4*
1 All causes of minor stoppages are analysed; all solutions implemented	1 Zero minor stoppages (unmanned operation possible)

Figure 5.9 OEE assessment: idling and minor stoppage losses

Level 1	*Level 2*
1 Equipment specifications not well understood 2 No speed standards (by product and machinery) 3 Wide speed variations across shifts/operators	1 Problems related to speed losses analysed: mechanical problems, quality problems 2 Tentative speed standards set and maintained by product 3 Speeds vary slightly
Level 3	*Level 4*
1 Necessary improvements being implemented 2 Speed is set by the product. Cause and effect relationship between the problem and the precision of the equipment 3 Small speed losses	1 Operation speed increased to design speed or beyond through equipment improvement 2 Final speed standards set and maintained by product 3 Zero speed losses

Figure 5.10 OEE assessment: speed losses

Level 1	*Level 2*
1 Chronic quality defect problems are neglected 2 Many reactive and unsuccessful remedial actions have been taken	1 Chronic quality problems quantified by: details of defect, frequency; volume lost 2 Losses categorized and reasons explained; preventive measures taken on trial and error basis
Level 3	*Level 4*
1 All causes of chronic quality defects analysed; all solutions implemented, conditions favourable 2 Automatic in-process detection of defects under study	1 Quality losses from 0.1% to zero

Figure 5.11 OEE assessment: quality defect and rework losses

Level 1	*Level 2*
1 Start-up losses not recognized, understood or recorded	1 Start-up losses understood in terms of breakdowns and changeovers 2 Start-up losses quantified and measured
Level 3	*Level 4*
1 Process stabilization dynamics understood and improvements implemented 2 Causes due to minor stops aligned with start-up losses	1 Start-up losses minimized through process control 2 Remedial actions on breakdowns, set-ups, minor stops and idling minimize start-up losses

Figure 5.12 OEE assessment: start-up losses

Loss Category	Improvement strategy examples
Breakdowns	Improve detection of conditions contributing to this, spot problems early.
Set-up losses	Identify in/outside work and organize/standardize. Identify unnecessary adjustments and eliminate.
Minor stops	Use P–M analysis. Cleaning will probably be a key factor.
Reduced speed	Identify speed, capability/capacity through experimentation. Speed up process to magnify design weaknesses. Use P–M analysis to identify contributory factors.
Quality losses	Classify causes and develop countermeasures, including standard methods to reduce human error.
Start-up losses	Establish key control parameters, minimize number of variables, define standard settings.

Figure 5.13 Reducing/eliminating the six losses

5.2 Condition cycle

Step 4 Critical assessment

The aim here is to assess the equipment production process and to agree the relative criticality of each element. This will enable priority to be allocated for the conditional appraisal, refurbishment, future asset care and improvement of those elements most likely to have an effect on overall equipment effectiveness.

The approach is to review the production process so that all members of the team understand (probably for the first time!) the mechanisms, controls, material processing and operating methods. Operators and maintainers must be involved in identifying the most critical parts of the process from their own perspective.

The important components and elements of the process, machine or equipment are identified: some typical examples are electrics, hydraulics, pneumatics, cooling systems and control systems. Each of these elements is assessed in terms of criteria such as the following:

- *Safety* If this component was in poor condition or failed, what would be the impact on safety due to increased risk of injury?
- *Availability* If this component was in poor condition or failed, what would be the impact on the availability of the equipment, including set-up and the need for readjustment of equipment settings?
- *Performance* What impact does this component have on the cycle time or processing capacity of the equipment when it is available to run?
- *Quality* If this component were in poor condition or failed, what impact would it have on product quality at start-up and/or during normal production?
- *Reliability* What impact does the frequency with which this component fails have on the overall criticality of the equipment?
- *Maintainability* What impact does this component have on the ease of maintaining or repairing the equipment?
- *Environment* If this component was in poor condition or failed, what would be the impact on the environment due to emissions, noise, fluid spills, dust, dirt, etc.?
- *Cost* If this component was in poor condition or failed, what would be the impact on total cost, including repair and lost production?
- *Total* The sum of the rankings for each component.

The significance of each of the criteria is assessed and allocated a score according to impact on the process: 1 = no impact, 2 = some impact, 3 = significant impact.

A typical matrix form for recording process elements and criteria scores is shown in Figure 5.14. The right-hand (totals) column enables priority to be applied to those elements most affected. This is further illustrated in Figures 5.15 and 5.16.

The main outputs from the critical assessment process are that it:

- starts the teamwork building between operators and maintainers;
- results in a fuller understanding of their equipment;
- provides a checklist for the condition appraisal;
- provides a focus for the future asset care;
- highlights weaknesses regarding operability, reliability, maintainability.

The critical assessment matrix provides the basis for understanding not just the most critical components but also those which contribute to special loss areas. For example, high scores on S, M and R indicate components which have a high impact on safety, are unreliable and difficult to maintain. A score of 6 or above on these three is an accident waiting to happen.

Other useful subsets include:

CRITICAL ASSESSMENT

Equipment description	1-3 Ranking as impact on:								
	S	A	P	Q	R	M	E	C	Total

Where S = Safety, A = Availability, P = Performance, Q = Quality
R = Reliability, M = Maintainability, E = Environment, C = Cost
1 = No impact, 2 = Some impact, 3 = Significant impact

Figure 5.14 Critical assessment matrix form

Overall equipment effectiveness	A, P and Q
Ease of use	P, Q and R
Maintainability	M, C and R
Environmental risk	E, M and R
Reliability	A, P and R

Revising those components with a high impact on quality is a good starting point for quality maintenance activities. Providing the assessment is applied consistently, it can also be used to establish basic maintenance strategies such as condition based (P = 2+) or run to failure (C = 1, M = 1, A = 3). These can then be refined as asset care routines are introduced and improved.

Step 5 Condition appraisal

The objective here is to make use of the same critical assessment elements and components in order to assess the condition of equipment and to identify the refurbishment programme necessary to restore the equipment to maximum effectiveness.

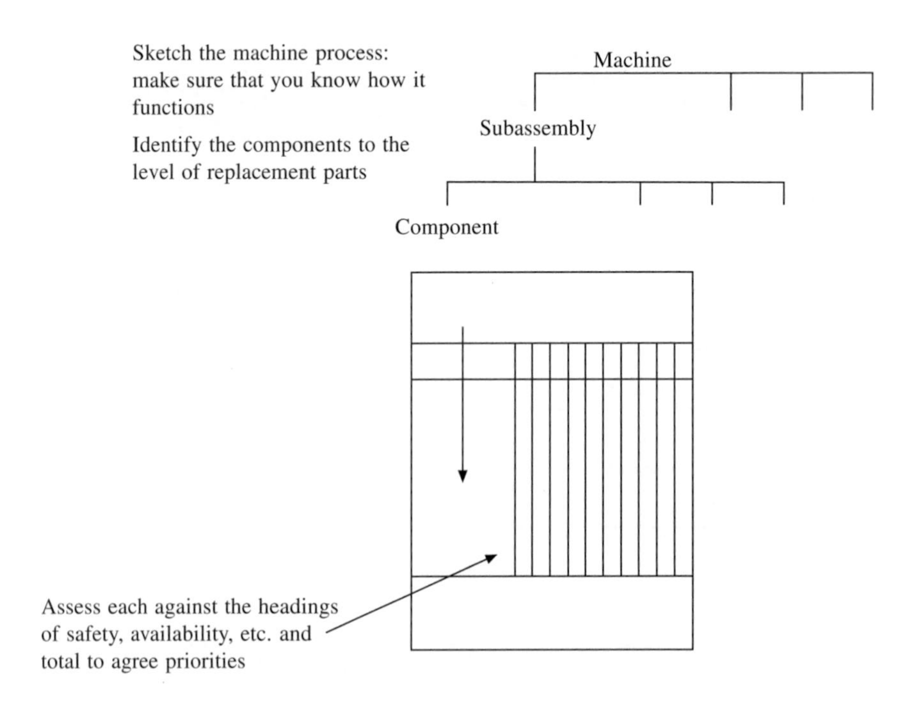

Figure 5.15 Stages in critical assessment

CRITICAL ASSESSMENT

Equipment description		Process: RH door hinge reinforcement 2 1-3 ranking as impact on: Welder								
	Component	S	A	P	Q	R	M	E	C	Tot
1	A-C Modicon 894145 plc	1	3	3	3	1	1	3	3	18
2	1/10 Modules DEP 2160	1	3	3	1	1	1	1	3	14
3	1/10 Modules DEP 216	1	3	3	1	1	1	1	3	14
4	Door SW RT 522586	1	3	3	1	1	1	1	3	14
5	Fluorescent Tube F20W-07/RS	1	1	1	3	1	1	1	1	10
6	Cooling Fans rS 509-068	3	2	1	1	1	3	3	1	15
7	Filters	1	2	2	1	1	1	2	1	11
8	Pushbutton C 2-BU06	1	2	1	1	1	1	1	1	9
9	2-B2-b N/C	1	2	1	1	1	1	1	1	9
10	2-b2-B101 N/C	1	2	1	1	1	1	1	1	9
11	Indicator 2-BU6	1	3	1	1	1	1	1	1	10
12	P Button SW 2-BU06	1	2	1	1	1	1	1	1	9
13	E/Stop Button	3	3	3	1	1	1	3	1	16
14										
15										
16										
17										
18										
19										
20										

Where S = Safety, A = Availability, P = Performance, Q = Quality, R = Reliability, M = Maintainability, E = Environment, C = Cost
1 = No impact, 2 = Some impact, 3 = Significant impact

Figure 5.16 Completed critical assessment matrix

Each heading will have been subdivided as necessary: for example, the electrical section may contain power supply, control panels, motors and lighting.

Under each of the subdivisions of the equipment being studied, four categories should be established:

- Satisfactory
- Broken down
- Needs attention now
- Needs attention later

An example of the outcome of a condition appraisal study is shown in Figure 5.17.

The key point of the condition appraisal is to put each square centimetre of the equipment under the microscope and assess whether its condition is 'as new' or 'as required'. Make sure also that you look inside the machine, so remove all panels. This is not just a broad, superficial look – on the contrary, it is being obsessive about attention to detail.

As such, the condition appraisal stage must include a deep clean of the equipment.

Step 6 Refurbishment

The objective of the refurbishment programme is to set up a repair and replacement plan, based on the condition appraisal, and indicating the resources needed. Getting the equipment back to an acceptable level is a prerequisite to the pursuit of ideal conditions.

The plan will provide a detailed summary of actions to be co-ordinated by the team, which will include:

- dates and timescales
- resources (labour, materials, time)
- cost estimates
- responsibilities
- control and feedback (management of change)

A typical summary table of refurbishment costs and man-hours required for a group of critical machines is shown in Figure 5.18.

The chart in Figure 5.19 gives details of action required on a specific item of equipment. It allocates responsibility for the various tasks and nominates individuals to carry out the work; it also embodies a simple visual indication of work progress.

The refurbishment programme is concerned not just with clearly identifiable repair work, but also with the many small weaknesses identified by the condition appraisal, including cleaning and CAN DO approach, such as missing bolts, leaks, temporary repairs and over/under-lubrication, and it highlights critical points for regular attention.

	Condition appraisal – Top sheet		
Machine No	VM56694	Description	RH front door
Date installed	20.07.91		Hinge re-
Commissioned	01.08.91		inforcement co
Warranty ends	01.08.92		welder
Location code	K19		
Plant priority			
Generic group	High	Marker	ESTIL
PO number	RH F/Door	Manufacturer	0766/01/00
Common	Assy	Serial No	
Equipment	0619862	Marker's No	
		Equipment Status	Operational
		Equipment Availability	Night & Day

General statement of reliability

General Statement

The access to the

Because of the SM

Also the welding

	Condition Appraisal		1 of 3				
Machine description							
Asset No:	Year of purchase		Appraisal by:				
Machine No:	Location:		Appraisal Date:				

Item No	Appraisal rating by sub asset	Not applicable	Satisfactory	Broken down	Needs attention now	Needs attention later
			'X' As required			
1	Electrical					
	A - Power supply to machine		×			
	B - Panel				×	
	C - Control/ls				×	
	D - Control circuits				×	
	E - Motors				×	
	F - Machine lighting	×				
	G - Trunking				×	
2	Mechanical					
	A - Spindle housings/Gearboxes				×	
	- Seals					×
	- Bearings					×
	- Gears		×			
	B - Slideways/Tables				×	
	- Workpiece				×	
	- Toolholder				×	
	C - Screws/Rams/Slined Shafts	×				
	D - Pneumatics				×	

Figure 5.17 Example of condition appraisal study

SUMMARY OF TOP 20 CRITICAL MACHINES –Refurbishment programme							
Asset No	Machine description	Est refurbishment cost			Man-hours work		
		Req'd now	Req'd later	Total	Req'd now	Req'd later	Total
873	Richmond Bed Drill	70	800	870	39	160	199
929	Devlieg 5 Pallet	16 135	6250	22 385	41	98	139
134	Maxicut	10	800	810	9	–	9
871	Snowgrinder 65/7–5	250	10 600	10 850	14	10	24
876	No 2 Mill	555	16 100	16 655	58	89	147
877	Radyne Bed Hardener	1260	990	2250	40	92	132
443	Voumard	30	80	110	5	8	13
477	Workmaster	2740	3850	6590	78	40	118
461	Naxos	245	35	280	10	14	24
652	Landis	370	–	370	24	32	56
847	Devlieg	190	2760	2950	24	122	146
848	Devlieg	60	17 760	17 820	17	212	229
879	No 4 Mill	85	–	85	11	40	51
858	Hydro 540	160	8270	8430	21	4	25
925	CNC 650 C-Axis I	3330	230	3560	28	40	68
926	CNC 650 C-Axis II	190	790	980	17	40	57
941	CNC 650	150	100	250	6	4	10
6	Snow Grinder 170/20	235	10 550	10 785	25	14	39
649	Snow Grinder 170/20	195	12 520	12 715	46	51	97
875	No 1 Mill	365	14 000	14 365	13	12	25
	Total	22 625	106 485	133 110	526	1082	1608

Figure 5.18 Refurbishment example for a group of machines; how costs can be spread

	Unit EA subs		Team LH Front Door ABC	TPM Refurbishment Chart		Manager G Booth R O'Neil	Supervisor F Robinson RB Seymour			Date 30.11.92
	Diversion 631		Operation CO_2 Welder							
	No	Item to be Refurbished	Action required	Responsibility	Completion Date	Champion	Check Date			Progress
Mech	1	Double actuating cylinder air feed	Repair/ replace	D Hudson	6.12.99	K Doyle				◑
Mech	2	Hose carrier chaffing	Relocate	R Edwards	6,12,99	K Doyle				◑
Mech	3	Slide mounting bolts	Replace/check tightness	C Taylor	6.12.99	I Howie				◑
Mech	4	Flexible gas supply pipe	Change from blue to orange	R Price	6.12.99	F Gebbington				◑
Elec	5	Control panel cooler fan	Resite	L Philips	6.12.99	M Cooper				◑
Elec	6	Control panel lights	Replace	M Cooper	6.12.99	T Foley				◑
Elec	7	Isolator	Identify	M Craig	6.12.99	I Lamen				◑
Elec	8	Solderoid valves	Identify	I Lamen	6.12.99	R Cassidy				◑
Elec	9	Rectifier cables	Tidy/re-route	C Weatherston	6.12.99	M Cooper				◑
Elec	10	Wire feed unit	Repair	L Philips	6.12.99	R Price				◑
Elec	11	Anti-spatter reservoir	Investigate why inoperative/ repair	R Price	6.12.99	D Carroll				◑
OPS	12	General machine	Clean/paint	I Howie	6.12.99	T Foley				◑

Fill in first segment when a refurbishment has been flagged | Fill in second segment when action is defined and responsibility is given | Fill in third segment when action is completed | Fill in fourth segment when refurbishment has been tested

CO_2 Mig Welding M/C:

Labour costs:

2 × 16 hours = 32 hours at £15.00	=	£480
1 × 13 hours = 13 hours at £14.00	=	£182
Total labour	=	**£662**
Parts:		
New seals to clamp cylinder	=	£15.00
6 PX leads	=	£60.00
Water flow gauge	=	£10.00
New air ducting	=	No cost
Water pressure gauge	=	£113.00
Total parts	=	**£198.00**
Total Mig Welding M/C	=	**£860.00**

Figure 5.19 Refurbishment example for a specific machine

Step 7 Asset care

Once refurbishment of an item of equipment has been carried out, a future asset care programme must be defined to ensure that the machine condition is maintained. It is therefore necessary to establish:

- cleaning and inspection routines
- checking and condition monitoring methods and routines
- planned preventive maintenance and service schedules

For each of these we must develop:

- improvements to make each task easier
- visual techniques to make each task obvious
- training to achieve consistency between shifts

It is important to distinguish between natural and accelerated deterioration. In the course of normal usage, *natural* deterioration will take place even though the machine is used properly. *Accelerated* deterioration arises from outside influences. These are *equipment-based,* i.e. failure to tackle the root causes of dust, dirt and contamination; and *operator-based,* i.e. failure to maintain basic conditions such as cleaning, lubricating and bolting, and also human operational errors.

Figures 5.20 and 5.21 show how the care of assets may be broken down into elements which reflect the first three steps of the condition cycle: criticality, condition and refurbishment. Figure 5.22 illustrates the relationship between operational and technical aspects of asset care. Some key points for consideration in asset care are shown in Figure 5.23.

The question of training is developed fully in Chapter 7, but some key approaches are illustrated in Table 5.1. A training schedule form is shown in Figure 5.24. This schedule is completed through a series of single-point, on-the-job lessons.

A practical example of daily cleaning and inspection is given in Figure 5.25. This shows the checks to be made in a MIG welding cell for each shift during the working week, and records all the daily checks made by the operators. A material usage chart first developed to highlight loss measurement

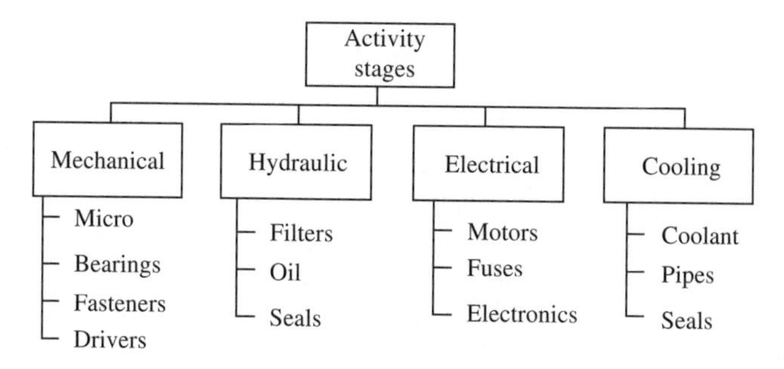

Figure 5.20 Stages in asset care

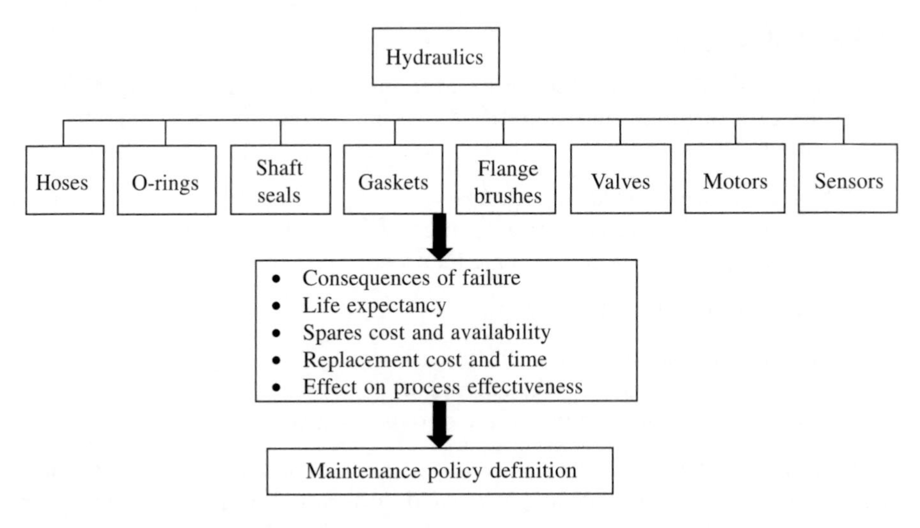

Figure 5.21 Breakdown of asset care for hydraulics maintenance

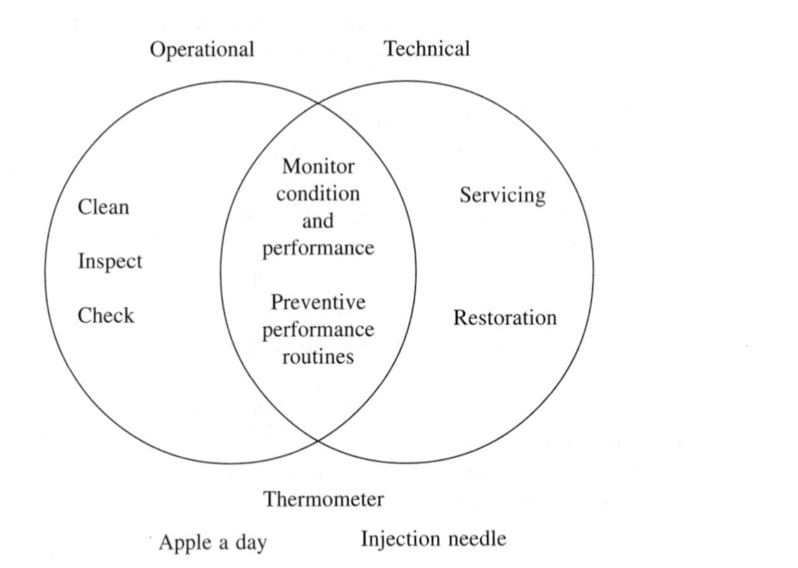

Figure 5.22 Relationship between operational and technical aspects of asset care (see also Figure 3.4)

For each task, make it:

- *Easy,* by simple improvements
- *Obvious,* using visual techniques
- *Consistent,* by effective training

Figure 5.23 Key points in asset care

Table 5.1 Role of training in asset care

Technique	*Learning points*	*Improvements*	*Training*
Cleaning	Accelerated deterioration Cleaning is inspection	Highlight vunerability Make it easy	Video
Inspection	What is the effect?	Provide tools/equipment	Single-point lessons
Checking	Check condition Check performance	Establish standards Establish parameters	Single-point lessons
Condition monitoring	What are the signs? Using our senses Using instruments	Make change obvious Make detection easy Provide tools/equipment	Instruction
Planned preventive maintenance	What is to be done? How do we do it? Who and when?	Make it accessible Make it maintainable Clear responsibility	Single-point lessons
Servicing	How do we manage it?	Clear instructions	

is now incorporated as a routine part of asset care in Figure 5.25. The key point is that the operators and maintainers have developed these asset care routines on the basis that 'If it's my idea, I will stick with it'.

The nine-step improvement plan provides a comprehensive analysis toolbox capable of reducing sporadic losses to zero, providing the appropriate infrastructure is in place. This infrastructure must include a continuous drive to reduce chronic losses by striving for optimum conditions. Not only does this keep people motivated to carry out the essential routine tasks, it provides progressively higher company competence to direct towards improved customer services.

The progressive implementation of front-line asset care and preventive maintenance provides the improvement zone partnership to deliver such improvement. If the nine-step improvement plan provides the answer to what is required, then the improvement zone implementation process provides the answer to how it is to be delivered. The how, where and when of this is part of management's role in 'creating the environment' for TPM (discussed in Chapter 8).

5.3 Problem prevention cycle

Step 8 Best practice routines

Following the restoration of equipment and development of asset care, the next step brings together all of the practices developed for operating,

TOTAL PRODUCTIVE MAINTENANCE PROGRAMME																
R/H FRONT DOORLINE MIG WELDING CELL																
T Foley *Team Leader.......*	OPERATOR TRAINING SCHEDULE															
Team member's name	Check air gauge	Clean M_1/c	Check CO_2 Wire	Change CO_2 Wire	Clamp check	Air & water leaks	Torch check	Top up fluid	Clean shroud	Check smog hog	Record & reset cycle counter	Check part levels	Parts on board	Scrap level	Rework level	
A Sefton																
B Dean																
P Lisle																
D Craven																
I Morris																
B Gallagher																
G Seddon																
M Lawton																
N Melling																
R Cross																
L Currie																
D Wells																
J Harwood																
S Morris																

Trained in procedures by maintenance — Carried out process — Competent in process — Able to train others

Figure 5.24 Training schedule form

TOTAL PRODUCTIVE MAINTENANCE PROGRAMME																			
R/H FRONT DOORLINE CO_2 MIG WELDING CELL																			
Date 28.9.99	DAILY CLEANING AND INSPECTION CHECKS																		
Notify Maintenance			×			×	×	×			×			×					
Work to be carried out ✓ If Ok × If Not Ok	Shift	Change CO_2 wire	Check pressure setting	Check table & tooling	Check CO_2 wire reel & spare	Check clamp head security	Check for air & water leaks	Top up anti spatter fluid	Check torch & harness security	Remove shroud and clean	Check smog hog light is on	Record & reset cycle counter	Check material letters	Check part on shadow board	Record scrap level	Record rework level			Signature
Monday	D/S	✓	✓	✓	✓	✓	✓	✓	✓	✓	✓	234	✓	×	✓	✓			BMG
	N/S	✓	✓	✓	✓	×	✓	✓	✓	×	✓	380	✓	✓	0	0			SC
Tuesday	D/S	✓	✓	✓	✓	✓	✓	✓	✓	✓	✓	328	✓	×	0	0			BMG
	N/S	✓	✓	✓	✓	✓	✓	✓	✓	✓	✓	200	✓	✓	0	0			DGW
Wednesday	D/S	✓	✓	✓	✓	✓	✓	✓	✓	✓	✓	326	✓	×	0	0			BMG
	N/S	✓	✓	✓	✓	✓	✓	✓	✓	✓	✓	345	×	×	0	0			SC
Thursday	D/S	✓	✓	✓	✓	✓	✓	✓	✓	✓	✓	103	✓	×	0	0			BMG
	N/S	✓	✓	✓	✓	✓	✓	✓	✓	✓	✓	386	×	✓	0	0			
Friday	D/S	✓	✓	✓	✓	✓	✓	✓	✓	✓	✓	187	✓	✓	0	0			BMG
	N/S																		
Saturday	D/S																		
	N/S																		

Figure 5.25 Example of daily cleaning and inspection checks

TOTAL PRODUCTIVE MAINTENANCE PROGRAMME														
R/H FRONT DOORLINE CO_2 MIG WELDING CELL														
	MATERIAL USAGE CHART													
Material	Week No 34	Week No 35	Week No 36	Week No 37	Week No 38	Week No 39	Week No 40	Week No 41	Week No 42	Week No 43	Week No 44	Week No 45	Week No 46	Week No 47
Welding wire														
Welding tip														
Shroud														
Anti-spatter spray														
Distilled water														

Figure 5.26 Material usage chart for example in Figure 5.25

maintaining and supporting the equipment, which are then *standardized* as the best practice routine across all shifts.

Figure 5.27 summarizes the relationship between standard operation, techniques for asset care and the right tools, spares, facilities and equipment.

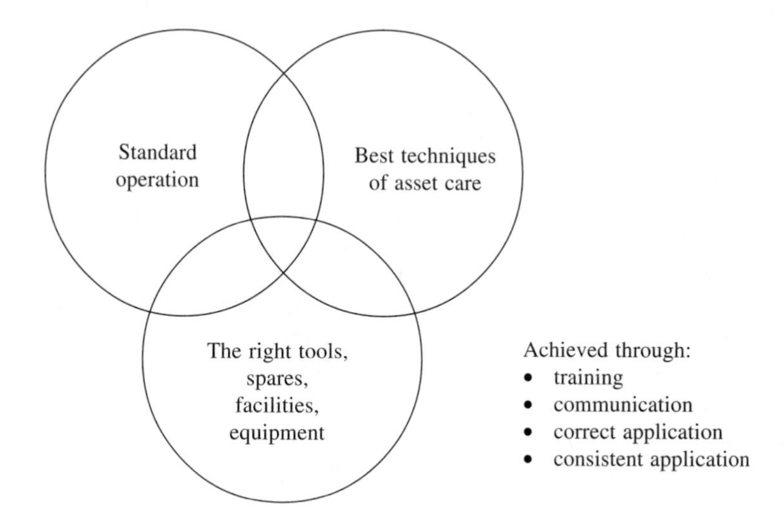

Figure 5.27 Best practice

Standard operation ensures:

- reduced chance of error and risk
- removal of performance irregularity
- elimination of poor operation as a cause of problems
- simplified training within and between shifts

When launching a pilot, we should consider:

- What is the best method of operating our pilot?
- Does the team agree?
- Does each shift agree?
- Do the key contacts agree?
- How do we train people to operate this method? (*single-point lessons*)
- How do we communicate this method to each shift? (*visual management*)
- How do we make it easy to do it right and difficult to do it wrong? (*improvements*)

For each piece of equipment we need to establish the best practice for:

- provision of tools
- provision of spares
- monitoring instruments
- outside contracts

- warranties
- technical help

We must therefore involve the key contacts.

In effect, the best practice routine is similar to your motor car handbook. It explains the best and correct way to operate, maintain and support the car. It gives the standard operation and asset care procedures.

Step 9 Problem solving

P–M analysis

Once the 'noise' of poor equipment care and different operative practices are addressed, the real technical problems can be revealed.

In seeking to solve the problems which lie behind the six big losses, TPM uses P–M analysis to address problems with mechanisms. This emphasizes the machine/human interface: there are *phenomena* which are *physical*, which cause *problems* which can be *prevented*; these are to do with *materials*, *machines*, *methods* and *manpower*.

These problems may have a single cause, multiple causes or a complex combination of causes. P–M analysis is concerned with pinpointing the causes, taking countermeasures and evolving best practice routines so that the problems are dealt with once and for all and fed back into an improved OEE.

On-the-job reality

This approach recognizes practical on-the-job steps as follows:

- *Cleaning is inspection* Operators are encouraged to look for opportunities to reduce accelerated deterioration and improve equipment design.
- *Detect problems and opportunities* Work with the team to systematically review problems and opportunities to achieve target performance:
  - Adopt a multi-stage approach
  - Make all aware of the problem and the opportunity
  - Observe the current situation and record
  - Define the problem and the conditions under which it occurs
  - Develop the optimum solution progressively
  - Try out new ideas first and check the results
  - Apply proven low-cost or no-cost solutions first
  - Implement ideas as soon as possible and refine quickly
  - Standardize best practice with all those involved
  - Monitor and review

Event review

For sporadic losses, the event review provides a simple way of trapping information, problem solving and developing countermeasures. The form shown in Figure 5.28 links five-why problem analysis with the five main countermeasures to sporadic losses. It is designed to promote post-event

EVENT REVIEW FORM

Plant No __________ Line No ___

Product ______________________

Opened date/shift	
Breakdown briefing	
Closed	

Ref	Brief description of fault and problem definition:		
How long ago could this defect have been detected?			
Why	Reason	Notes/Potential countermeasure	
1			
2			
3			
Short term actions			
Priority for review			
Proposed countermeasure(s) to prevent reoccurrence			Status
Modification to operator/maintainer practice/training			
IMPROVED Routine Activities			⊕
Check/prediction			⊕
Planned maintenance			⊕
Component modification			⊕
Other			⊕

Agreed Planned Implemented (inc SPL) Confirmed

Figure 5.28 Event review format

discussion between operators and maintainers. It also uses *status wheels* to report progress and provide a record to review the event of reoccurrence.

In Figure 5.29 we have completed an event review form using the overhead projector analogy discussed in Chapter 3.

Recurring problems

In order to resolve problems and prevent recurrence, knowledge and

EVENT REVIEW FORM

Opened Date/Shift	14/12/98
Breakdown Briefing	6–2 pm shift
Closed	

Plant No *O/head Projector* **Line No** *1*

Product *View Foils*

Ref *106*	Brief description of fault and problem definition: *Lamp bulbs keep blowing*	
How long ago could this defect have been detected? *Probably 1 month plus*		
	Reason	**Notes/Potential Countermeasure**
Why bulb blows 1	*Overheating*	*Uprated bulb to higher watts – still happens*
Overheating 2	*Air flow insufficient*	*No way of telling. Fan is working OK. Uprate fan?*
Why A/F insuff 3	*Filter blocked*	*Source of contamination*
Why filter block 4	*Not cleaned*	*Daily asset care needed*
Short term actions	*Clean filter*	
Priority for review	*Top*	

Proposed countermeasure(s) to prevent reoccurrence		**Status**
Modification to operator/maintainer practice/training	*See below*	
IMPROVED Routine activities	*Clean filter once a week on Friday Shift – Operator*	
Check/prediction	*Set up thermometer to check temperature in the box. Colour code red/green. Record each day*	
Planned maintenance	*Overhaul and check fan rating every six months*	
Component modification	*Set up thermometer. Shadow board for bulbs*	
OTHER	*SPL needs for OPS recleaning and changing filter + temperature reading + bulb changing*	

Figure 5.29 Event review example

understanding is the key to training operators to be equipment-conscious. Some examples, checklists and techniques are given below.

Overheating, vibration and leakage are problems which will constantly arise and, unless tackled and eliminated once and for all, will continue to contribute to breakdown losses. Tables 5.2–5.5 offer approaches to these problems.

A structured approach to set-up reduction is necessary. Table 5.6 draws attention to all the points which must be looked at and evaluated. An indication

Table 5.2 Problem solving: leakages

	Cause	*Remedy*
1	Excessive vibration	Cure cause
2	Unabsorbent mountings	Refit new mountings
3	Insufficient mountings or supports	Fit extra
4	Wrong grade/type component fitted	Fit correct grade
5	Poor fitting	Refit correctly
6	Overheating	Seek and cure cause
7	Technical ignorance/innocence	Retrain
8	Material breakdown	Replace

Vibration is one of the major causes of fittings or fixings working loose and giving rise to leaks. Other items contribute, such as poor fitting, or overheating, which causes seals first to bake and then crack.

To identify leaks:

- In the case of liquids: puddles will form
- In the case of gases: noise, smell or bubbles when tested with soapy water.

Table 5.3 Problem solving: overheating

	Cause	*Remedy*
1	Excessive lubrication	Remove excess
2	Incorrect lubricant	Replace with correct
3	Lubrication failure/contamination	Check cause and remedy
4	Low lubricant level	Top up
5	Poor fitting	Refit correctly
6	Excessive speed above standard	Reduce speed to standard
7	Overloading	Reduce loading
8	Blockages in system	Clean and flush system
9	Excessive pipe lengths or joints	Redesign system

Table 5.4 Problem solving: overheating and lubrication

When *overheating* can be attributed to a lubricating problem, it is always best policy to remove all lubricant and replace with new after the problem has been cured. Lubricant which has overheated starts to break down and will not perform as it should.
Identification of overheating:

- *Visual* Items that have overheated will discolour or give off smoke.
- *Smell* In many cases overheated items give off fumes which can be smelt.
- *Touch* By touching items suspected of overheating one can tell, but caution must be exercised in the first instance. A hand held close to the item will indicate whether it would burn if touched.
- *Electrical/ Visual* Many items of equipment have built-in temperature-sensing devices and these should be monitored regularly. An awareness of the significance of the temperature readings is essential.

Table 5.5 Problem solving: vibration

	Cause	*Remedy*
1	Out of balance	Correct or replace
2	Bent shafts	Straighten or replace
3	Poor surface finish	Rework surface
4	Loose nuts and bolts	Tighten
5	Insecure clips	Secure clips
6	Insufficient mountings	Get extra added
7	Too rigid mountings	Get softer ones
8	Slip stick	Lubricate
9	Incorrect grade lubricant	Clean and replace
10	Worn bearings	Replace
11	Excessive speed above standard	Reduce speed to standard

Some of the remedies will require a skilled maintenance fitter. Others can be carried out by the operator with some training (items 4, 5, 8, 9 and 11).

Vibration is identified by sight, touch or noise increase.

of the importance of tackling adjustment is given by the percentage figures based on hard experience shown in Table 5.7.

Set-up and adjustment are so important in the drive towards reduced losses, better equipment effectiveness and ultimately world-class manufacture. Shigeo Shingo, the guru of Single-Minute Exchange Die (SMED), states the following in his book *A Revolution in Manufacturing: the SMED System*: 'Every machine set-up can be reduced by 75%'.

What a challenge for Western companies! The SMED approach uses a derivative of the Deming circle:

Focus	Set-up video
Analyse	Pareto, ergonomics
Develop	Script, simulate, agree
Execute	Train, measure, honour, empower

In the SMED system, success is subject to certain conditions:

• *An attitude*	The team wants to score.
• *An empowerment*	The team has a budget.
• *An involvement*	Management is part of the team.
• *A commitment*	Management sets the target.
• *A philosophy*	Step-by-step improvement.

Moreover, the SMED approach suggests that there are characteristics common to all set-ups:

- Prepare, position, adjust, store away
- Internal and external activities
- From last good product to first good product

Table 5.6 Factors in reduction of set-up and adjustment time

External set-up Preparations	• Tools (types, quantities) • Locations • Position • Workplace organization and housekeeping • Preparation procedure	• Don't search • Don't move • Don't use
Preparation of ancillary equipment	• Check jigs • Measuring instruments • Preheating dies • Presetting	
Internal set-up	• Standardize work procedures • Allocate work • Evaluate effectiveness of work • Parallel operations • Simplify work • Number of personnel • Simplify assembly • Assembly/integration • Elimination	• Eliminate redundant procedures • Reduce basic operation
Dies and jigs	• Clamping methods • Reduce number of clamping parts • Shapes of dies and jigs: consider mechanisms • Use intermediary jigs • Standardize dies and jigs • Use common dies and jigs • Weight • Separate functions and methods • Interchangeability	• Make it easy
Adjustment	• Precision of jigs • Precision of equipment • Set reference surfaces • Measurement methods • Simplification methods • Standardize adjustment procedures • Quantification • Selection • Use gauges • Separate out interdependent adjustments • Optimize conditions	• Eliminate adjustment

Table 5.7 Adjustments as a percentage of total set-up time

Preparation of materials, jigs, tools and fittings	20%
Removal and attachment of jigs, tools and dies	20%
Centring, dimensioning	10%
Trial processing, adjustment	50%

Figure 5.30 shows the three steps towards a cumulative reduction of 75 per cent to 95 per cent in set-up time in the SMED system. A graphical representation of the reductions achieved is shown in Figure 5.31.

TPM develops six conceptual steps for analysing adjustment operations.

1 *Purpose* What function is apparently served by adjustment?
2 *Current rationale* Why is adjustment needed at present?
3 *Method* How is the adjustment performed?

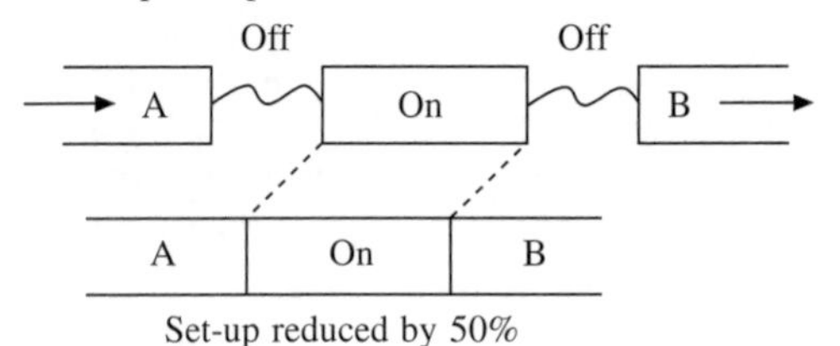

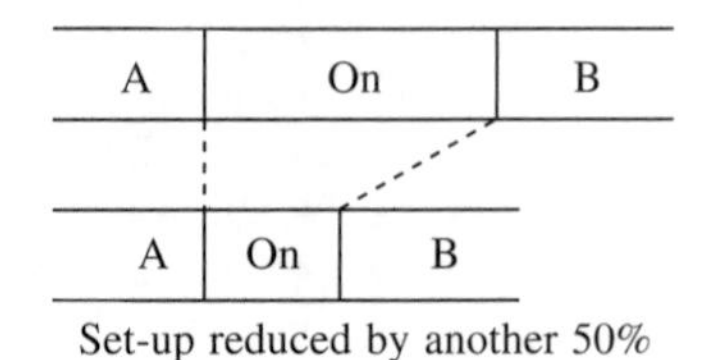

Step 3: Minimize external activities and continue reducing internal activities

A	On	B

A	On	B

Cumulative reduction between 75% and 95%!

Figure 5.30 SMED steps to reducing set-up time

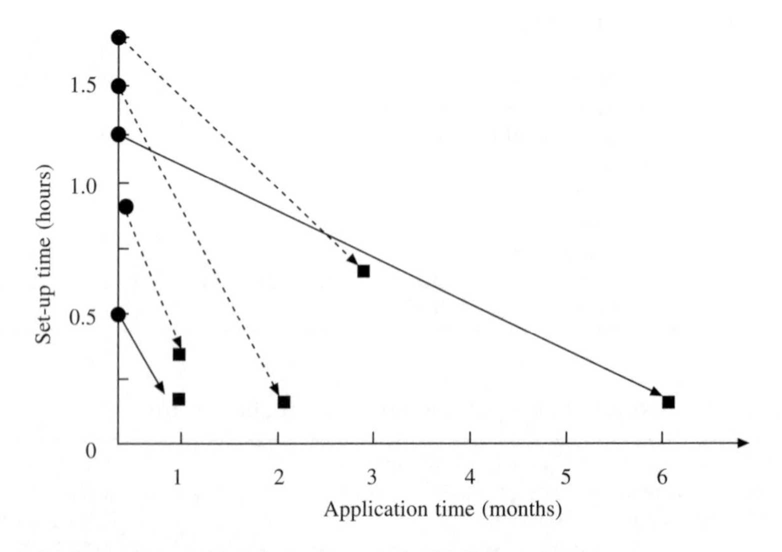

Figure 5.31 Set-up times reduced significantly by SMED approach

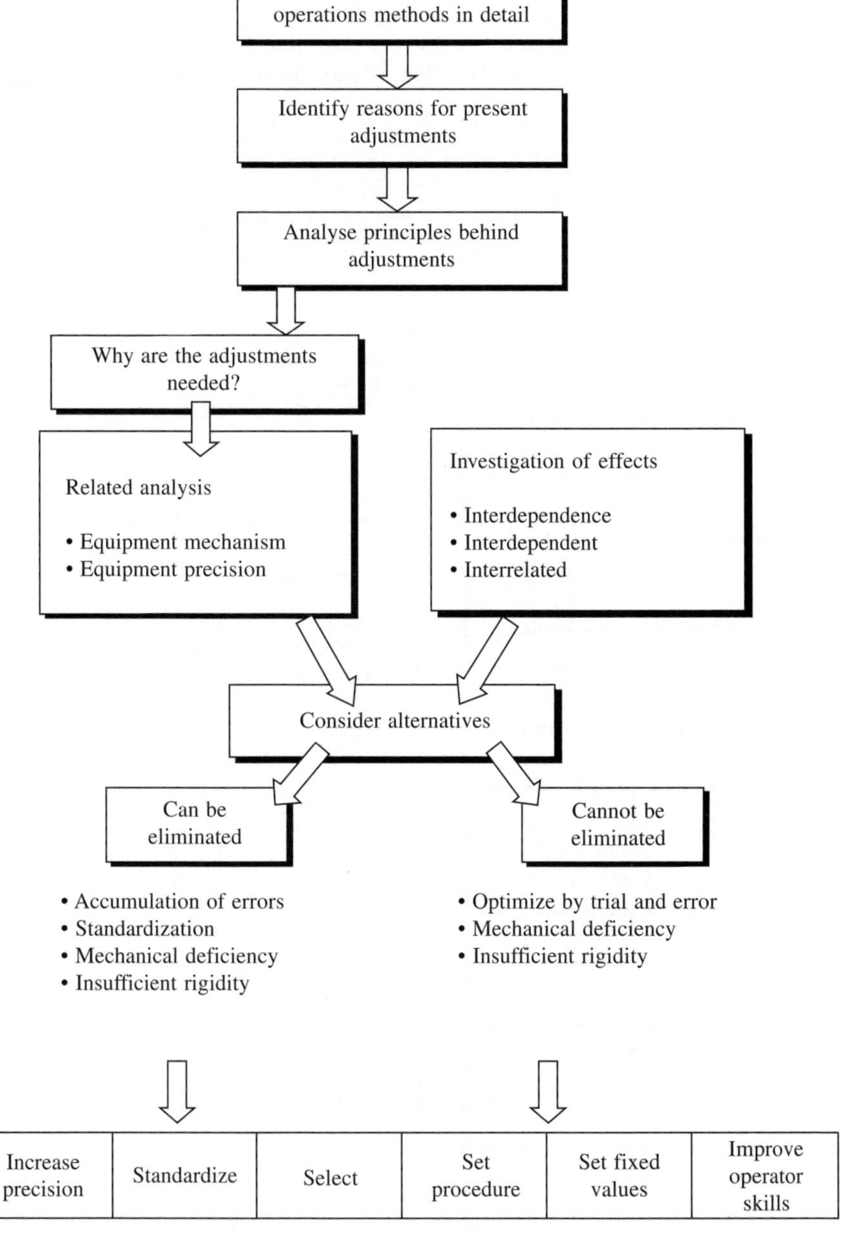

Figure 5.32 Analysis of adjustment operations

4 *Principles* What is the true function of the adjustment operation as a whole?
5 *Causal factors* What conditions create the need for adjustment?
6 *Alternatives* What improvements will eliminate the need for adjustment?

Figure 5.32 provides a clear visual presentation of the TPM approach to analysing adjustment operations leading to minimization of losses. Figure 5.33 reviews progressively the process from an analysis of the present position right through to achieving optimal conditions. Wherever possible, make use of video: it is a very powerful analysis tool.

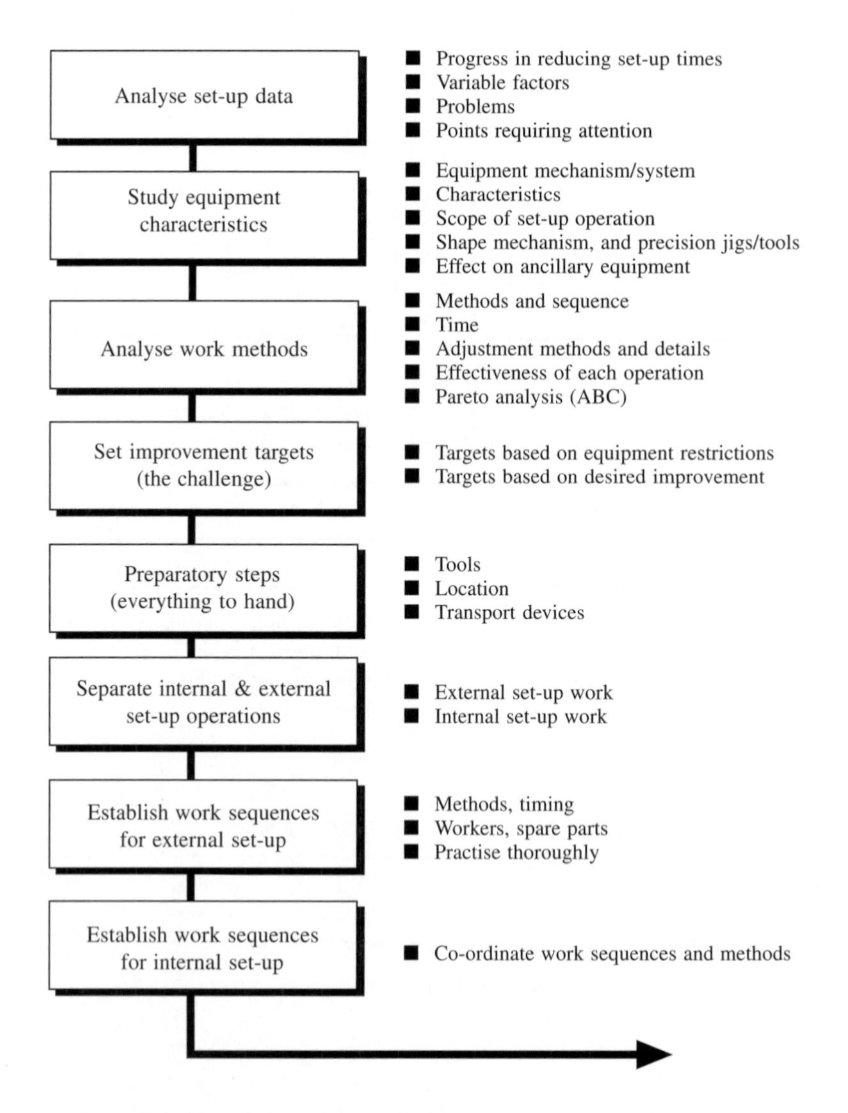

Figure 5.33 Process of improving set-up and adjustment

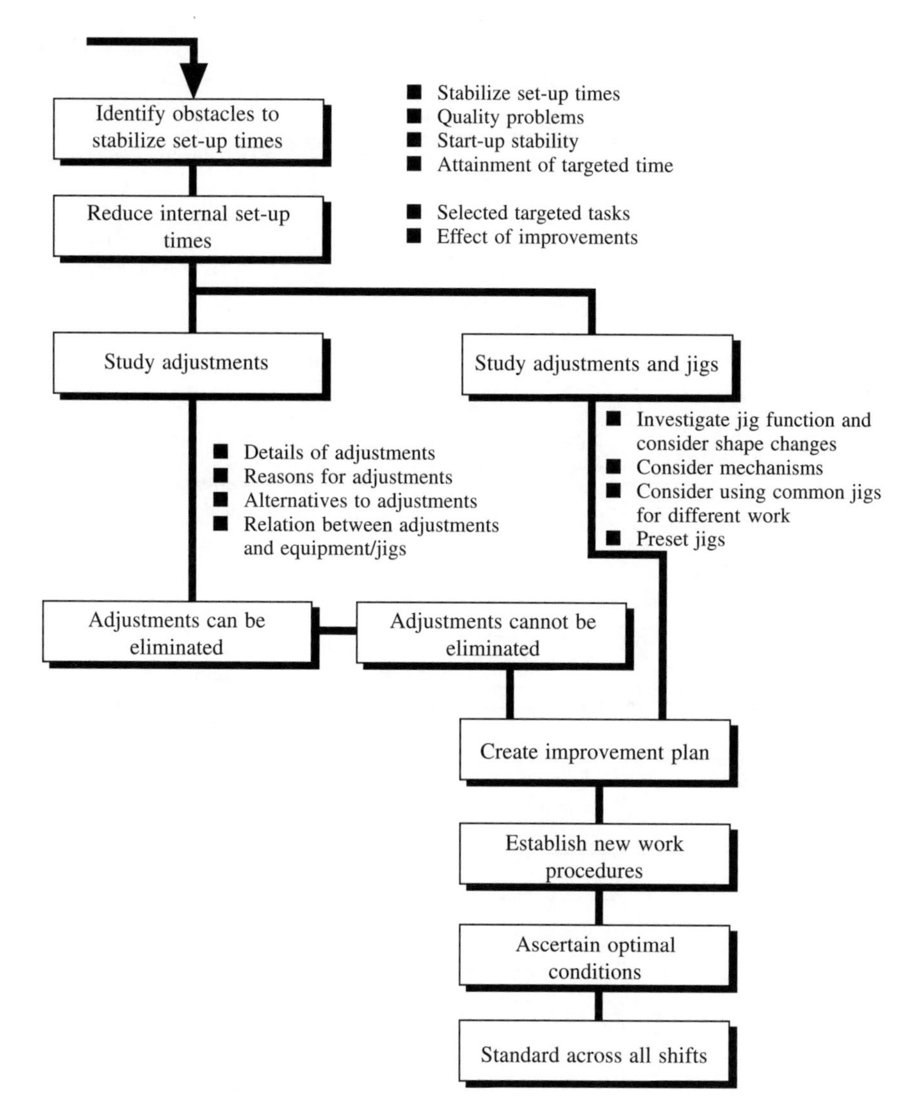

Figure 5.33 continued

Figure 5.34 provides a decision-tree structure to help eliminate reasons for running at reduced speeds. Table 5.8 provides a checklist of ideas for developing approaches to increase speeds.

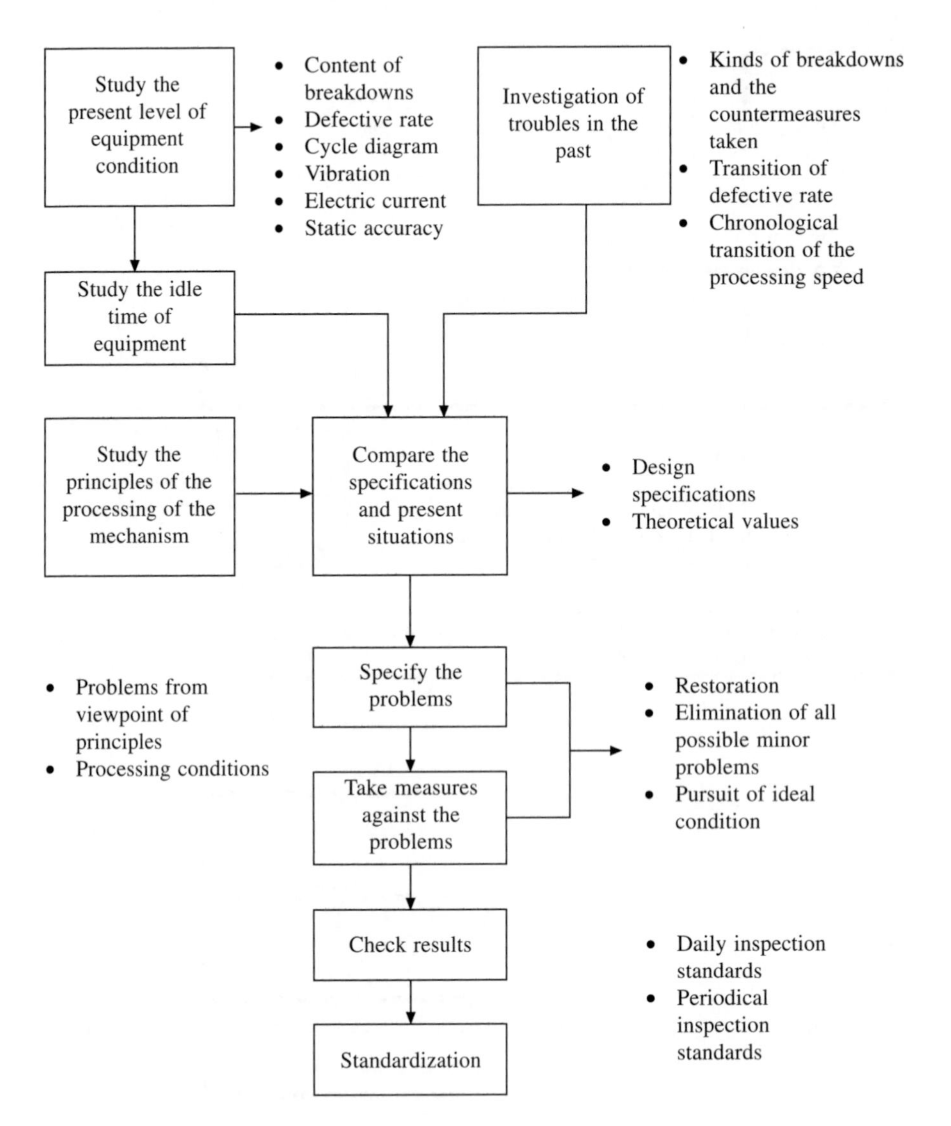

Figure 5.34 Countermeasures for speed losses

Table 5.8 Strategies for increasing speed

Determine present levels	• Speed • Bottleneck processes • Downtime, frequency of stoppages • Conditions producing defects
Check differences between specification and present situation	• What are the specifications? • Difference between standard speed and present speed • Difference in speeds for different products
Investigate past problems	• Has the speed ever been increased? • Types of problems • Measures taken to deal with past problems • Trends in defect ratios • Differences in similar equipment
Investigate processing theories and principles	• Problems related to processing theories and principles • Machining conditions • Processing conditions • Theoretical values
Investigate mechanisms	• Mechanisms • Rated output and load ratio • Investigate stress • Revolving parts • Investigate specification of each part
Investigate present situation	• Processing time per operation (cycle diagram) • Loss time (idling times) • Check precision of each part • Check using five senses

6 Applying the TPM improvement plan

6.1 Training context

The following example is based on a TPM improvement plan training exercise as part of a four-day TPM facilitator training course. Approximately 70 per cent of the course content is putting the theory of TPM into practice on a *live* TPM pilot piece of equipment. This practical focus is so that the facilitators are experiencing over four days what they will be coaching their own TPM teams over the twelve to twenty weeks of a typical TPM pilot project, the process of which is described in Chapter 7.

The output of this particular exercise is based on a one-hour feedback presentation which the five budding TPM facilitators made after spending two and a half days assimilating and using the nine-step TPM improvement plan. The following sections 6.2 to 6.16 inclusive are the content of their presentation.

6.2 Team brief

A core team is undertaking a pilot TPM project. The team is made up of:

- Three production personnel (one per shift)
- Two maintenance personnel (one electrical, one mechanical)

The company – Merlin Gun Technology – is planning to introduce TPM across the site (200 personnel).

The pilot aims to develop a practical, model example of equipment operating under TPM. This will support the roll-out of TPM. It will also highlight those issues which need to be overcome to achieve a successful implementation. Specifically, the team will:

- assess the critical elements of the equipment;
- identify what refurbishment is required to put the equipment into good condition;
- develop a practical, model example of equipment in good condition;
- develop an asset care and history recording process;
- identify the main problem areas and develop solutions;
- establish the current level of overall equipment effectiveness and set targets for improvement;

- produce an implementation plan to improve the equipment reliability.

Background

Merlin Gun Technology has the following characteristics:

- The company makes welding guns and welding tips.
- Most of the volume is in welding tips.
- The company experiences pressure from customers to produce in small batches.
- The company is expanding into further export markets.
- Department 50 is recognized as the main bottleneck.

The company organization is shown in Figure 6.1.

Department 50

Department 50 makes the most popular welding tips. It has always been the bottleneck department.

The department produces basically three types of tip (Figure 6.2):

- The 5020, a flat-head tip produced in two operations
- The 5031, a tip with one angled face produced by three operations
- The 5042, a pointed tip with two angled faces produced in four operations

There are three machines:

- The L101 computer numerically controlled (CNC) lathe
- The M201 CNC Bridgeport Interact milling machine
- The M202 Denford Easimill 3 milling machine

The machine operating data are shown in Figure 6.3. It should be noted that the times have been developed by the planners based on experience. The machines should be capable of the following cycle times:

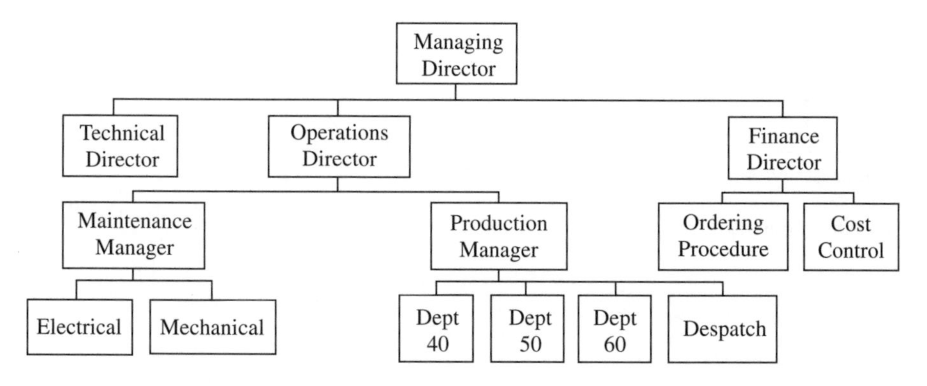

Figure 6.1 Merlin Gun Technology: organization

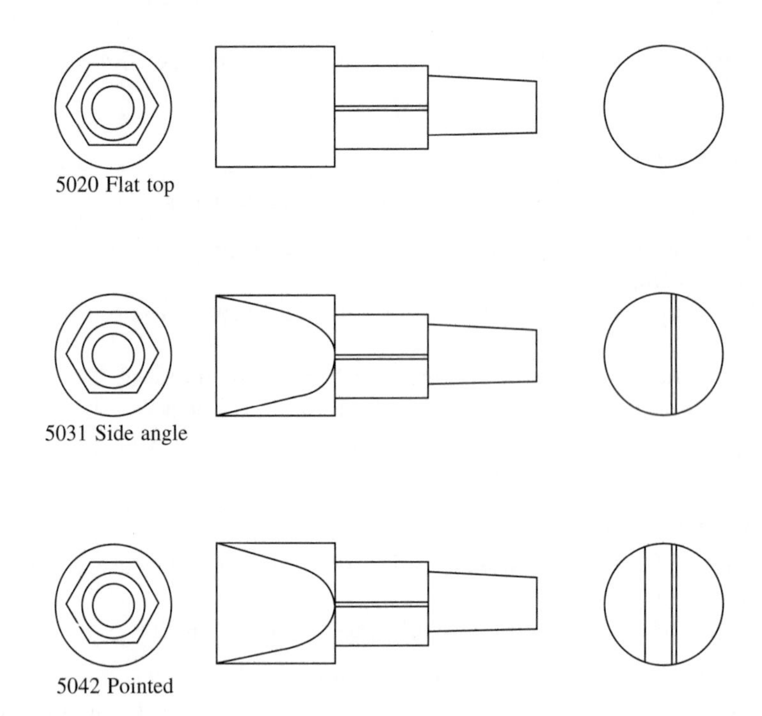

Figure 6.2 Department 50 products

Part	Forecast	Operations				Total cycle time (mins)
		L101	M201	M202		
				RH angle	LH angle	
5020	690	✓	✓			13.50
5031	200	✓	✓	✓		29.50
5042	65	✓	✓	✓	✓	45.50
Output per week		955	850	330		
Scheduled running time (min)		7200	4800	5200		
Scheduled time per cycle (min)		7.5	6.00	16.00	16.00	

Figure 6.3 Department 50 machine operating data

- L101: 4.12 min including loading
- M201: 3 min including loading
- M202: 8.5 min including loading

As can be seen from the data in Figure 6.3, the lathe is running three shifts five days per week; the miller M201 is running two shifts, five days; and the miller M202 is running two shifts plus overtime. Some cover is provided for the machines during the shifts, but there are still times when the machines cannot be run. These planned stoppages are one of the reasons for the difference between them.

There are some tool changes. Sometimes this is done by maintenance outside production time. Usually some production time is lost owing to tool breakages.

There seem to be occasional quality problems where much of the output has to be reworked to remove burrs. However, mostly there are few quality problems.

Department 50's financial data are shown in Figure 6.4. The company's shopfloor logistics are given in Figure 6.5.

6.3 TPM presentation and plan

The equipment chosen for the pilot project is the M201 milling machine.

The TPM presentation for the M201 team had the following headings:

- Introduction
- Plan
- Equipment description
- Equipment history
- OEE assessment
- The six losses
- Criticality assessment
- Condition appraisal
- Refurbishment
- Future asset care
- Best practice routines
- Problems/improvements
- Implementation plan
- Concluding remarks

The schedule for the project is shown in Figure 6.6.

As discussed in earlier chapters, the TPM improvement plan has the following three cycles and nine steps.

Measurement cycle

- Decide what to record and monitor
- Decide the OEE measures: best of best, world class
- Assess the six losses

	£'000	£'000	
Profit forecast			
Sales		1322.36	
Materials	550.00		
Consumables	2.72		
Inventory adjustment	6.13		
Material costs		558.85	
Direct labour	138.15		
Pensions	23.14		
Holiday pay	13.27		
Labour costs		174.56	
Production	188.55		
Depreciation	37.78		
Production overheads		226.33	
Technical	61.26		
Administration	104.15		
Selling	77.26		
Finance	33.35		
Other overheads		276.02	22%
Total costs		1235.76	
Profit		86.60	7%

Cost apportionment

Part	*Production forecast*	*Material cost/piece*	*Cycle time* (min)	*Machine time* (h)	*Fixed costs/product* (£)	*Fixed costs/piece* (£)
5020	45 540	9.40	13.5	10 246.50	228 046	5.01
5031	10 120	9.40	29.5	4 975.67	110 739	10.94
5042	3 680	9.40	45.5	2 790.67	62 109	16.88
	59 340			18 012.84	400 894	

Revenue forecast

Part	*Total cost/piece* (£)	*Operational margin* (%)	*Selling price* (£)	*Total revenue* (£)
5020	14.41	25	18.01	820 153
5031	20.34	50	30.51	308 800
5042	26.28	100	52.55	193 402
				1 322 355

Figure 6.4 Department 50 financial data

Condition cycle

- Critically assess the equipment
- Carry out an appraisal of its condition

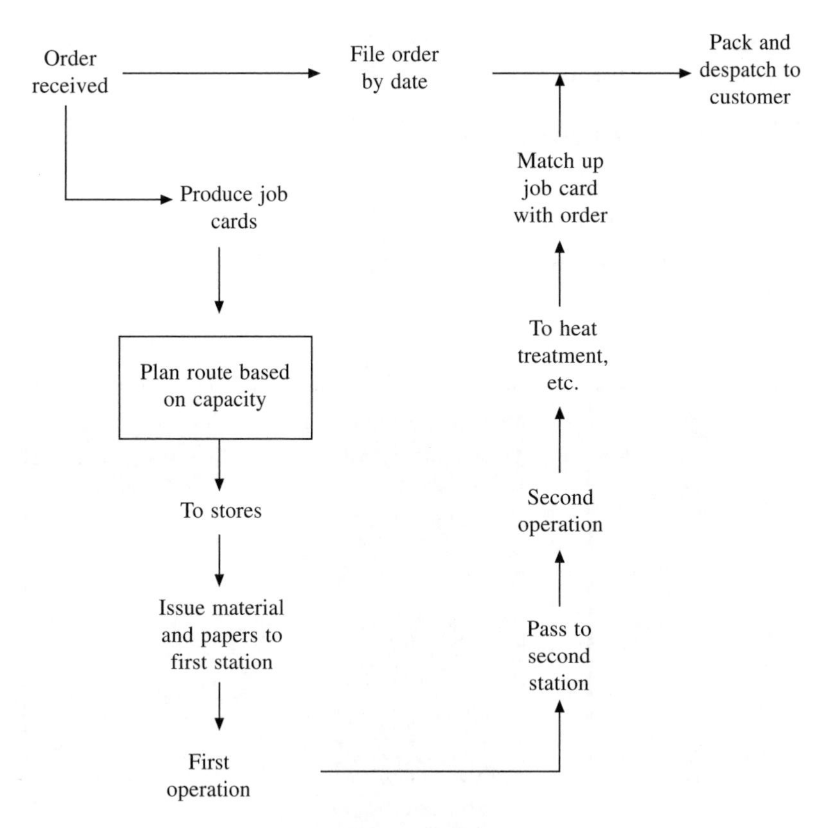

Figure 6.5 Merlin Gun Technology: shopfloor logistics

Activity	Tuesday pm	Wednesday am	Wednesday pm	Thursday am
Plan				
Equipment description				
Collect equipment history				
Define OEE				
Assess 6 losses				
Criticality assessment				
Condition appraisal				
Refurbishment plan				
Future asset care				
Best practice				
Problems/improvements				
Implementation plan				
Presentation preparation				

Figure 6.6 Miller M201 project management plan

- Decide on the refurbishment programme
- Determine the future asset care regime

Problem prevention cycle

- Agree on best practice routines
- Achieve improvement through problem solving and prevention

For convenience, the nine-step TPM improvement plan is repeated here as Figure 6.7.

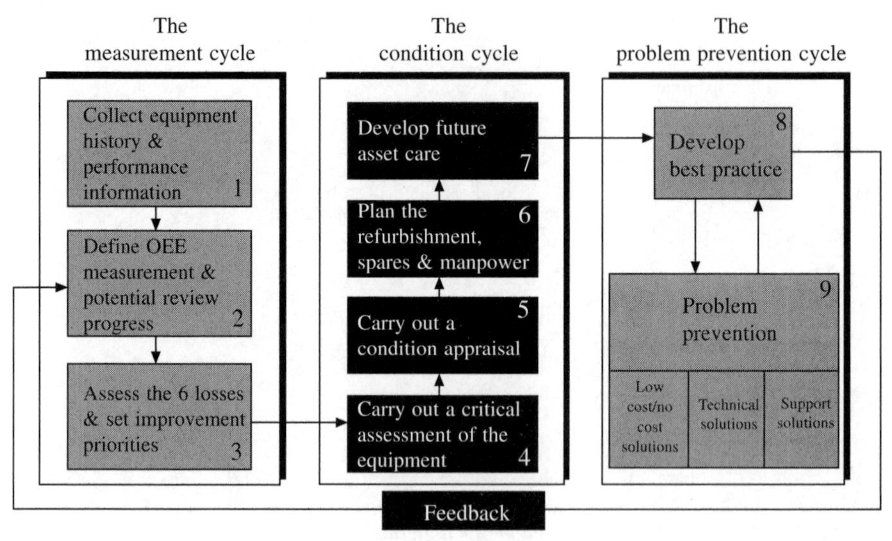

Figure 6.7 The TPM improvement plan

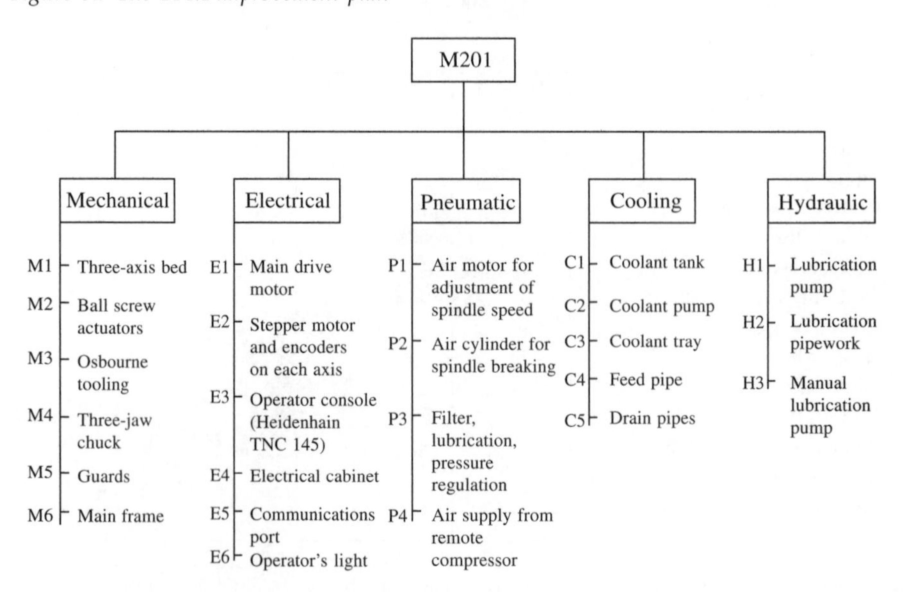

Figure 6.8 Miller M201 major components

6.4 Equipment description

As a first step, some of the main components of the M201 miller are described in Figure 6.8 and illustrated in Figure 6.9.

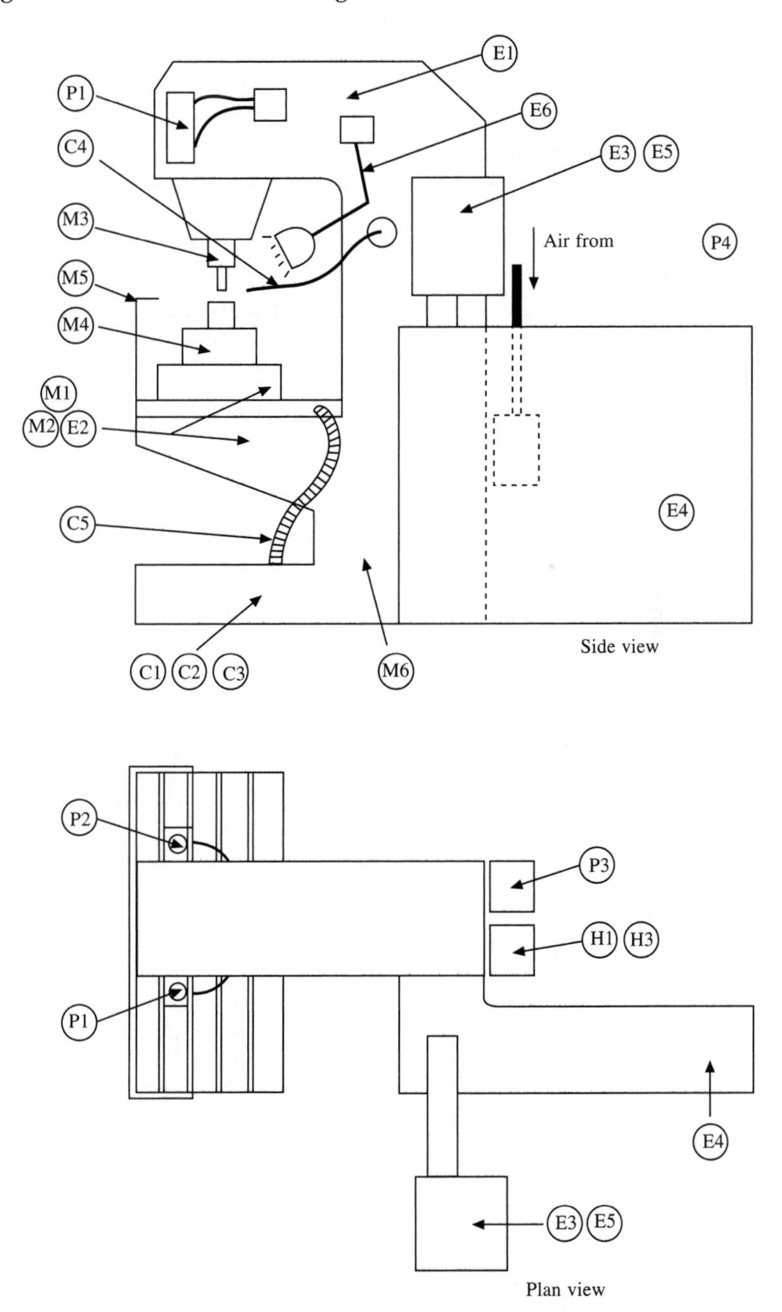

Figure 6.9 Miller M201 component diagram (see Figure 6.8 for key)

The performance data for the M201 are given in Table 6.1. A three-week equipment history is provided by Table 6.2.

The M201 operation cycle is shown in Figure 6.10. The operations layout for Department 50 is given in Figure 6.11.

Table 6.1 Miller M201 performance data

Maximum spindle speed	3750 rpm
Maximum axis feed rate	1 m/min
Current spindle speed	2000 rpm
Current x and y axes feed rates	0.5 m/min
Current z axis feed rate	25 mm/min
Time for first cut of hexagonal	33 s
Time for subsequent hexagonal cuts	15 s each
Total number of subsequent cuts	4
Total time for all cuts	93 s
z axis travel per cut	2 mm
z axis total travel	10 mm

Table 6.2 Miller M201 equipment history

Cycle time *3.00* min

Week no	*Day, based on 1440 min*	*Planned stoppages (min)*	*Planned availability (min)*	*Unplanned stoppages (min)*	*Changeover (min)*	*Uptime (min)*	*Completed cycles*	*Rework*	*First time OK*
15	M	530	910	120		790	187		187
	T	530	910	153	20	737	185		185
	W	530	910	96		814	124	13	111
	Th	530	910	132	20	759	209		209
	Fr	530	910	129	13	769	175	13	162
16	M	530	910	90		820	151	118	33
	T	530	910	42	9	859	228		228
	W	530	910	42	9	859	228		228
	Th	530	910	99	20	791	121		121
	F	530	910	50		860	201		201
17	M	530	910	27	32	851	223		223
	T	530	910	115	12	783	203	51	152
	W	530	910	34	24	852	214		214
	Th	380	1060	356	5	700	132		132
	F	530	910	46	17	847	236	76	160
									2546

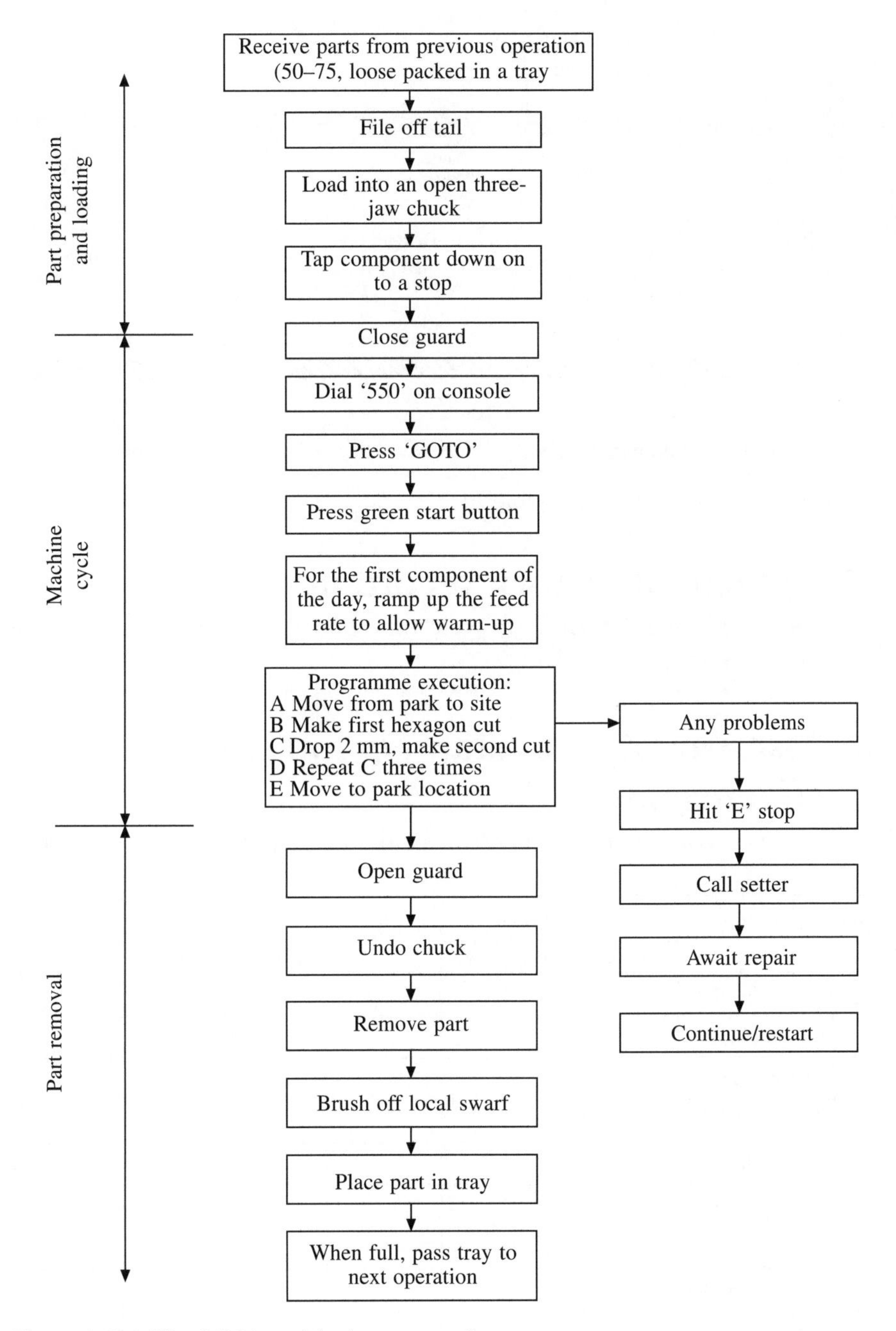

Figure 6.10 Miller M201 machine/operator cycle

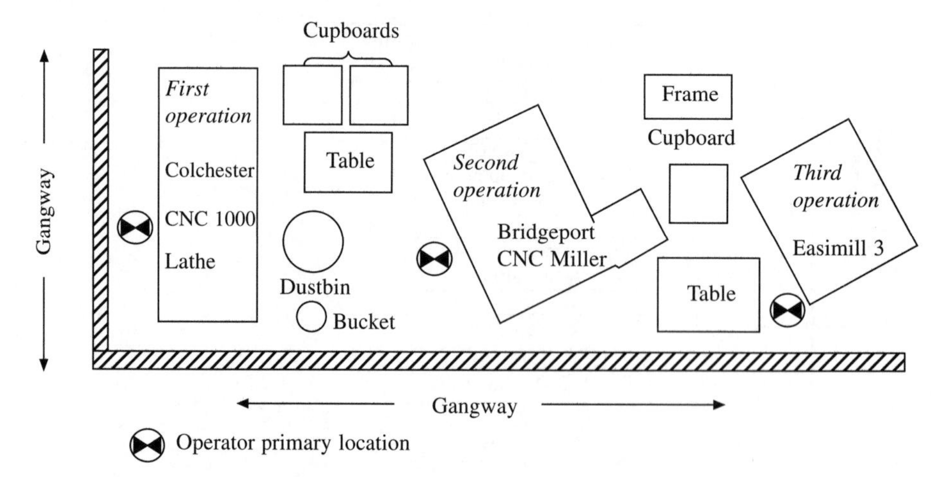

Figure 6.11 Department 50 operations layout

6.5 Equipment history recording

Newly designed record forms for cycle time and downtime are shown in Figures 6.12 and 6.13 respectively.

6.6 Assessment of overall equipment effectiveness

The OEE is given by the relation OEE = availability × performance × quality.

Week no		Day	Components			Changeover time	Stoppages		Signature
			Produced	Rejects	Total		Planned	Unplanned	
	D N	Monday							
	D N	Tuesday							
	D N	Wednesday							
	D N	Thursday							
	D N	Friday							
	D N	Monday							
	D N	Tuesday							
	D N	Wednesday							
	D N	Thursday							
	D N	Firday							
Totals						(min)	(min)	(min)	

Figure 6.12 Miller M201 equipment history record: operations

Inc no	Date	Time	Equipment	Fault	Reason	Remedy

Figure 6.13 Miller 201 equipment history record: downtime and faults

Example OEE calculation

The following example uses the data for the Tuesday of week 17 on the equipment history record of Table 6.2.

Availability

$$\text{availability} = \frac{\text{uptime}}{\text{planned availability}}$$

$$\text{uptime} = \text{planned availability} - \text{downtime}$$

$$\text{downtime} = \text{unplanned stoppages} + \text{changeovers} = 115 + 12 = 127 \text{ min}$$

Therefore

$$\text{uptime} = 910 - 127 = 783 \text{ min}$$

$$\text{availability} = \frac{783}{910} = 86.0\%$$

Performance

$$\text{performance} = \frac{\text{completed cycles}}{\text{planned cycles}}$$

$$\text{completed cycles} = 203$$

$$\text{planned cycles} = \frac{\text{uptime}}{\text{standard cycle time}} = \frac{783}{3} = 261$$

Therefore

$$\text{performance} = \frac{203}{261} = 77.7\%$$

Quality
(RFT = right first time)

$$\text{quality} = \frac{\text{components (RFT)}}{\text{completed cycles}}$$

$$\text{components (RFT)} = 152$$

$$\text{completed cycles} = 203$$

Therefore

$$\text{quality} = \frac{152}{203} = 74.9\%$$

OEE = 86.0% × 77.7% × 74.9% = 50%

A summary of the OEE values for the equipment history provided (Table 6.2) is given in Table 6.3. A simple graph of the OEE possibilities is shown in Figure 6.14.

Cost/benefit analysis

The cost/benefit analysis is based on the additional units that can be produced per week for each 1 per cent improvement in OEE.

Table 6.3 Miller M201 OEE summary

Date		*Availability*	*Performance*	*Quality*	*OEE*
15	M	86.8	71.0	100*	61.6
	T	81.0	75.3	100	61.0
	W	89.5	45.7	89.5	36.6
	Th	83.4	82.6	100	68.9
	F	84.5	68.3	92.6	53.4
16	M	90.1	55.2	21.9	10.9
	T	94.4	79.6	100	75.1
	W	94.4	79.6	100	75.1
	Th	86.9	45.9	100	39.9
	F	94.5*	70.1	100	66.2
17	M	93.5	78.6	100	73.5
	T	86.0	77.7	74.9	50.0
	W	93.6	75.3	100	70.5
	Th	66.0	56.6	100	37.4
	F	93.1	83.6*	67.8	52.8
Average		87.6%	69.9%	90.3%	55.3%
*Best of best		94.5%	83.6%	100%	79.0%

Difference between best of best and average: 24%

$\left(\text{A real improvement potential of } \frac{24}{55} = 44\%\right)$

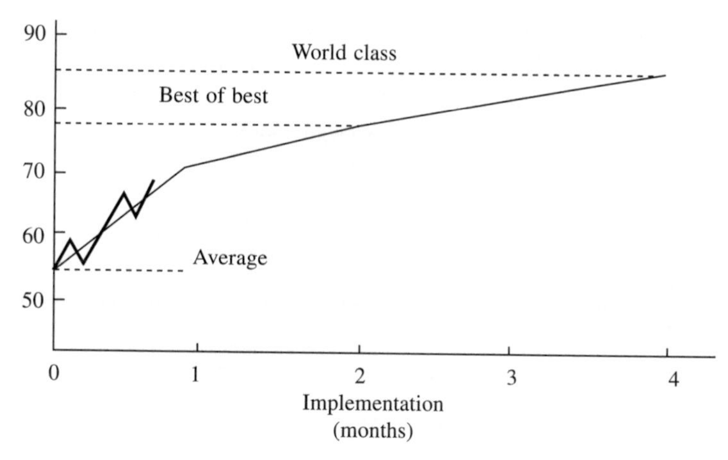

Figure 6.14 Miller M201 OEE comparison

$$\text{Units per week for each 1\% improvement in OEE} = \frac{\text{total components (RFT)}}{\text{average \% OEE} \times 3 \text{ weeks}}$$

$$= \frac{2546}{55 \times 3} = 15.43$$

Thus the benefit of increasing the average OEE up to the best of best OEE (approximately 79 – 55 = 24%) is equivalent to an extra 370 units per week.

6.7 Assessment of the six losses

Following an initial visit to the machine and a 'brainstorming' session with the operator and maintainer, the problems identified for the M201 are as follows:

- abnormal operation
- electrical power loss
- vibration
- no air supply
- tooling performance affected
- long cycle time
- initial start-up procedure
- excessive component loading time
- slideway damage
- operational safety
- no reference documentation

These problems and issues have been allocated to the six losses in Figure 6.15. Part of the loss assessment record is shown in Table 6.4.

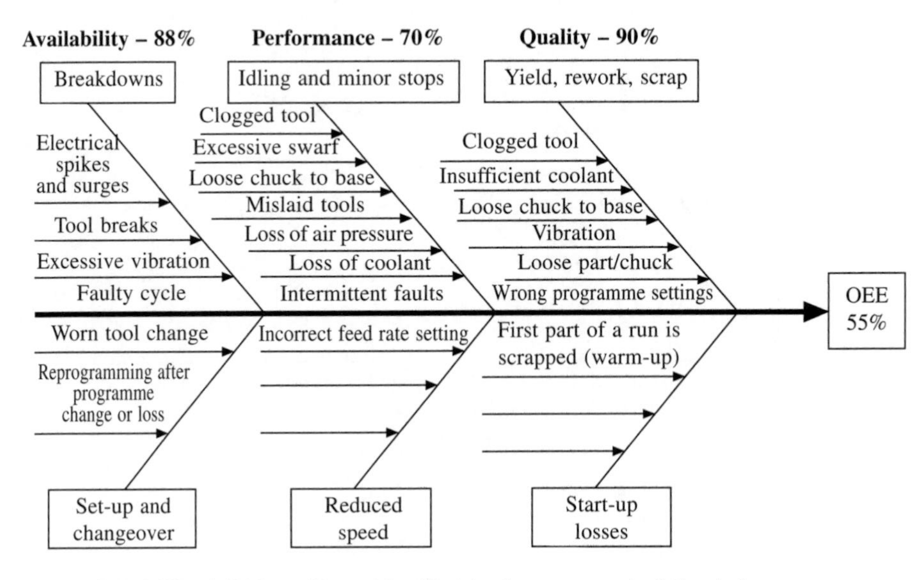

Figure 6.15 Miller M201: problems identified in the assessment of the six losses

6.8 Critical assessment

Following the initial six losses assessment and brainstorming session, the team now has a clearer understanding of the main components of Machine 201 and a critical assessment has been completed as shown in Figure 6.16. Reference to this Figure shows that the control computer and compressor have the highest ranking, closely followed by the cutting tool, the dead plate ball screw and the auto lube kit.

Other points to note are that the cutting tool (8 points out of a maximum of 9) has the biggest potential impact on the OEE, as does the work piece chuck and the auto services kit.

Where safety is a 3 and reliability a 3 on the same component, one can conclude that this may be an 'accident waiting to happen' (the cutting tool and control computer).

6.9 Condition appraisal

A completed condition appraisal form is shown as Figure 6.17.

6.10 Refurbishment programme

A study of the refurbishment requirements indicated the following:

- Tasks during machine operation 8 (20 hours)
- Tasks requiring downtime 12 (34 hours)
- Total tasks 19 (54 hours)

Table 6.4 Miller M201 loss assessment record

Loss type	*Item*	*Availability*	*Performance*	*Quality*	*Associated problems/issues*	*Impacts*
Breakdown	Electrical surge/spike	•		◎	• Lost/corrupted program requires setter to reload and test • Scrapped part • Possible damage to equipment	• Immediate stop • Additional start-up cycle • Program and inspection tests
Breakdown	Tool break	•		◎	• Scrapped part • New tool to be fitted • Setter needed to restart	• Immediate stop • Purchase new tool
Idling and minor stops	Clogged tool		•	◎	• Affects surface quality • Reduces cutting efficiency, hence increased load on machine, accelerated wear • Operator needs to clean tool	• Reduced performance • Progressive build-up affecting quality, not identified by operator: scrap parts
Idling and minor stops	Loose chuck		•	◎	• Affects part quality (surface and size) • Operator needed to retighten	• Reduced performance • Progressive loosening affects quality, not identified by operator: scrap parts
Yield/rework	Insufficient coolant	◎		•	• Affects part dimension (growth) • Reduces tool life • Increases load on machine	• Reduced quality • Reduced availability due to tool change

• = primary impact
◎ = secondary impact

CRITICAL ASSESSMENT

	M201 Bridgeport								
Equipment Description	1–3 Ranking as impact on:								
	S	A	P	Q	R	M	E	C	Total
Servo Motor Spindle	1	1	3	1	1	1	1	1	10
Spindle Motor	1	1	3	1	1	1	1	1	10
Spindle Motor	1	1	3	2	1	3	1	3	15
Spindle SPD Change Motor	1	1	2	1	1	1	1	1	9
Spindle Air Brake	3	1	3	1	1	2	1	1	13
Cutting Tool	3	3	2	3	3	1	1	1	17
Spindle Brg/Slides	1	1	1	3	1	3	1	3	14
Servo Motor BEP	1	1	3	1	1	1	1	1	10
Work Piece Chuck	3	2	3	3	1	1	1	1	15
Dead Plate Ball Screw	1	1	3	3	1	3	1	3	16
Bed BLS/Slides	1	1	2	3	1	3	1	3	15
Control Computer	3	1	3	1	3	3	1	3	18
Elec Box	3	1	3	1	1	2	1	1	13
Suds Pump	1	2	1	2	1	1	3	1	12
Auto Lube Kit	1	3	3	2	3	1	2	1	16
Air Serv Kit	2	3	3	1	1	1	2	1	14
Compressor	2	2	3	1	3	3	1	3	18
Guards	3	1	1	1	1	1	1	1	10
Cleaning Kit	2	1	1	1	1	1	3	1	11
Gauging Tool	1	3	1	3	1	1	1	1	12

Where S = Safety
A = Availability
P = Performance
Q = Quality
R = Reliability
M = Maintainability
E = Environment
C = Cost

1 = No impact
2 = Some impact
3 = Significant impact

Figure 6.16 Critical assessment

The costs of this programme are expected to be as follows:

- Labour costs £790
- Material costs £330
- Total costs £1120

The major refurbishment tasks are:

- Replace slide blanket.
- Replace spindle gasket oil seal and bearing.
- Design workplace stop.
- Investigate vibration problem and cure.

6.11 Asset care

A preferred spares listing for the M201 miller is shown in Figure 6.18. Schedules for checking and monitoring and for daily cleaning and inspection are shown in Figures 6.19 and 6.20.

		Condition Appraisal – Top Sheet	
Machine No:	*M201*	Description:	*Interact CNC*
Date Installed:			*Milling machine*
Commissioned:			
Warranty Ends;			
Location Code:		Maker:	*Bridgeport*
Plant Priority:		Manufacturer Serial No:	
General Group:		Equipment Status:	
P.O. Number:		Equipment Availability:	
Common Equipment;			

General Statement of Reliability

Generally regarded by the operator as a reliable machine. The four most significant reliability issues are:

Electrical failure due to surges/spikes
Vibration problems
Air supply failures/low pressure
Running out of coolant (when used)

General Statement of Maintainability

There is no planned maintenance for this machine. The operator does not participate in any maintenance activity. Access for maintenance is severely hindered by the layout, i.e. tables and cupboards close to the machine.

Figure 6.17 Miller M201 condition appraisal record

		Condition Appraisal – Sheet 1 of 4			
Machine Description:		*M201, Bridgeport CNC, Milling Machine*			
Asset No:	Year of purchase:	Appraisal by:			
Machine No:	Location:	Appraisal Date:			
Item No	Appraisal Rating by Sub Asset	Satisfactory	Broken Down	Needs Attention Now	Needs Attention Later
		X as required			
1	Electrical				
	A – Power Supply to Machine			x	
	B – Panel			x	
	C – Control				x
	D – Control Circuits			x	
	E – Motors	x			
	F – Machine Lighting			x	
	G – Other				
2	Mechanical				
	A – Spindle Housings/Gearboxes			x	
	Seals			x	
	Bearings	x			
	Gears	x			
	B – Slideways/Tables			x	
	Workplace			x	
	Toolholder			x	
	C – Screws/Rams/Splined Shafts			x	
	D – Pneumatics			x	

Figure 6.17 (Contd)

	Condition Appraisal – Sheet 2 of 4				
		Satisfactory	Broken Down	Needs Attention Now	Needs Attention Later
		X as required			
2	Mechanical (continued)				
	E – Coolant System			x	
	F – Guards			x	
	G – Other				
3	Working Area				
	A – Layout			x	
	B – Hazards			x	
	C – Storage				x

Figure 6.17 (Contd)

	Condition Appraisal – Sheet 3 of 4
Asset No: *M201* Location	Description: *Bridgeport Milling M/c*

Sub Asset Generic Group

Denote condition as one of the following:

S = Satisfactory
B/D = Broken Down
NAN = Needs Attention Now
NAL = Needs Attention Later

Generic Group	Problem Found	Condition
1A	Electrical system susceptible to spikes	
*1B	Open hole on top surface – swarf/water	NAN
1D	System is temperamental at start-up	NAN
1F	Causes operator headaches	NAN
1G	Cables and panels covered in swarf	NAN
2A	Oil leak on spindle housing	
*2B	Slide protection blanket torn/holed	NAN
*2D	Air supply pipe too long – trip hazard	NAN
*2E	Kinked pipe	NAN
2E	Coolant tray not secure	
2E	Coolant tray damaged	
2F	Guard components not secure	NAN
2F	Compliance with current regulations?	NAN
2G	Vibration	NAN

Figure 6.17 (Contd)

	Condition Appraisal – Sheet 4 of 4	
Asset No: .... Location	Description:	
Generic Group	Problem Found	Condition
3A	Cabinets/table/bins restrict access	NAN
3B	Table motion creates nip point	NAN
3B	Wet floor – slip hazard (roof leak)	NAN
3B	Swarf everywhere, inc floor	NAN
3C	No defined location for parts, incoming, outcoming or scrap	NAN
3C	No defined location for brush, hammer chuck key, spare tools, file	NAN
3C	Setter's cabinets in disarray	NAN

Figure 6.17 (Contd)

6.12 Best practice routines

The key areas for attention and where best practice routines will need to be developed are as follows:

- *Asset care* Cleaning; monitoring; planned maintenance
- *Correct operation* Clear instructions; easy to operate; understand process
- *Good support* Maintenance and operator work together. Additional support

		Preferred Spares Listing	
Machine No: M201		Description: Bridgeport	
Location Machine Shop		Milling m/c	
Critical – ■	Dedicated – ■	Consumable – ■	
Description	Qty	Part No	Supplier/Type
Gaskets	6		Bridgeport
Gaitors	6		,,
Slide covers	1		,,
Swarf brushes	12		Local supply
Safety glasses	12		,,
Gloves	12		,,
Coats	12		,,
Guard spares	1		Manufacturers
Coolant tray	1		Bridgeport
Tool bits (various)	20		Local supply
Lub oil (lts)	200		,,
Coolant (paraffin) (lts)	50		,,
Coolant pipes/clips (mts)	10		,,
Fuses	12		,,
Builds	12		,,
Paint (lts)	5		,,
Lubricator	5		,,
Moisture trap	6		,,

Figure 6.18 Miller M201 asset care: spares listing

from accounts, production, design, purchase and planning (the key contacts)

- *Inspection* Operator's responsibility
- *Training* Operator and maintenance

An operator training plan is drawn up at Figure 6.21 which will be supported by highly visual single-point lessons.

Frequency 1 = per shift 1a = per day 2 = per week S = start E = end	SHIFT	S1 Oil reservoir/levels	S1 Air lubricator inspect top-up	S1 Moisture trap drain	E1 Motor temperature	S1 Pipelines air/oil/coolant	S1 Vice security alignment	S1 Table height	S1a Warm-up cycle and emergency stop	S1 Coolant levels	S1 Air pressure	E2 Vibration analysis motors and table	S1a Electrical isolators	E2 Maintenance dept survey/inspections	E2 Weekly clean routine oil slides and traverse gear			Signature
Monday	D																	
	N																	
Tuesday	D																	
	N																	
Wednesday	D																	
	N																	
Thursday	D																	
	N																	
Friday	D																	
	N																	
Saturday	D																	
	N																	
Inform maintenance					X	X			X		X	X	X		X			

Figure 6.19 Miller M201 checking and monitoring record

Frequency 1 = each component 2 = 2 per shift 3 = per shift S = start of shift E = end of shift	SHIFT	2 Clean M/C remove swarf	E3 Empty swarf tray and clean filter	E3 Brush and clean area around M/C	E3 Clean M/C table	E3 Inspect and clean M/C head	E3 Guards	S3 Clean and inspect tool condition	1 Tool cleaning	1 Vice and 3-jaw chuck	Signature
Monday	D/S										
	N/S										
Tuesday	D/S										
	N/S										
Wednesday	D/S										
	N/S										
Thursday	D/S										
	N/S										
Friday	D/S										
	N/S										
Saturday	D/S										
	N/S										
Notify maintenance					X	X			X		

Figure 6.20 Miller M201 daily cleaning and inspection record

6.13 Problem solving and improvements

An example of a problem-solving document is shown in Figure 6.22. The main improvements identified, and their effects on availability, performance and quality, are shown in Table 6.5.

6.14 Implementation

The proposed implementation programme is shown in Figure 6.23.

Team Leader: Jenny

Member's name															Date of completion
John															
Richard															
Martin															
Alex															
	Cleaning routines	Inspections	M/C monitoring	Vibration analysis	Safety training	Introduction to TPM	Tool changes	Sequence control	M/C set-up	Basic maintenance techniques	Fault finding	Housekeeping	Quality control	Data recording	1 × □ Trained in procedures 2 × □ Carried out process 3 × □ Competent in process 4 × □ Able to train others

Figure 6.21 Miller M201 operator training plan

TPM PROBLEM SOLVING DOCUMENT			
Machine/Part Name	*M201*	Team Leader:	
Problem Raised By:		Date:	*7/9/*
1. Problem Statement (Specific)	*Excessive cycle time resulting in low output*		
2. Clarification of problem (by whom, where, when and how) plus cost/benefit opportunity *Clarified with operator on 7/9/* *Discussed opportunities to reduce cycle time without sacrificing product quality.* *Cost to modify programme = 1 hour set-up and 1 hour test/inspect* *Benefit Standard cycle time reduced from 3 to 2 seconds = extra 152 units per week (a 10% OEE improvement)*			
3. Problem cause (brainstorm and fishbone, then list) 8/8 Caused by 1: *Shallow depth of cut* Caused by 2: *Movement of table in z-plane* Caused by 3: *Relative position of tooling start position and workpiece* Caused by 4: *Need to manually reset programme to start cycle due to multiple programmes in memory*			
4. Clarification of root causes (Do the causes explain the problem?) • *Single 10 mm cut is practical rather than 5 × 2 mm cuts* • *If single cut implemented, table height can be fixed via programme* • *Reduction of tool travel shortens cycle time* • *Programme location/multiple programmes prevent auto reset function being utilized, thus manually reset for each cycle*			
5. Countermeasure (actions required to resolve causes (s)) Temporary– Permanent • *Modify programme to single cutting cycle from 5 cuts* • *Set traverse table at constant height in programme* • *Reprogramme tool start position close to workpiece* • *Store single programme in memory and locate at line ϕ to enable auto reset statement to be used*			
6. Confirmation of countermeasure (Have actions cleared problem?) • *Countermeasures awaiting implementation*			
7. Feedback (Who else needs to know?) • *Miller Supervisor, Maintenance Manager, TPM Facilitator, nominate team for TPM Excellence Award (Sept)*			

Figure 6.22 Miller M201 problem-solving document

Table 6.5 Miller M201 improvements

	Benefit for		
	Availability	*Performance*	*Quality*
Reduce cycle time		high	
Programme modifications		high	
Operator to load programmes	low		
Use of coolant	low		medium
Operator to change tooling	medium		
Reposition air supply service unit	low		
Define start-up procedure		high	low
New slideway cover design	low		
Obtain copies of manual/drawings	low		
Replace with new design guards	low		
Operator training programme	high	high	high

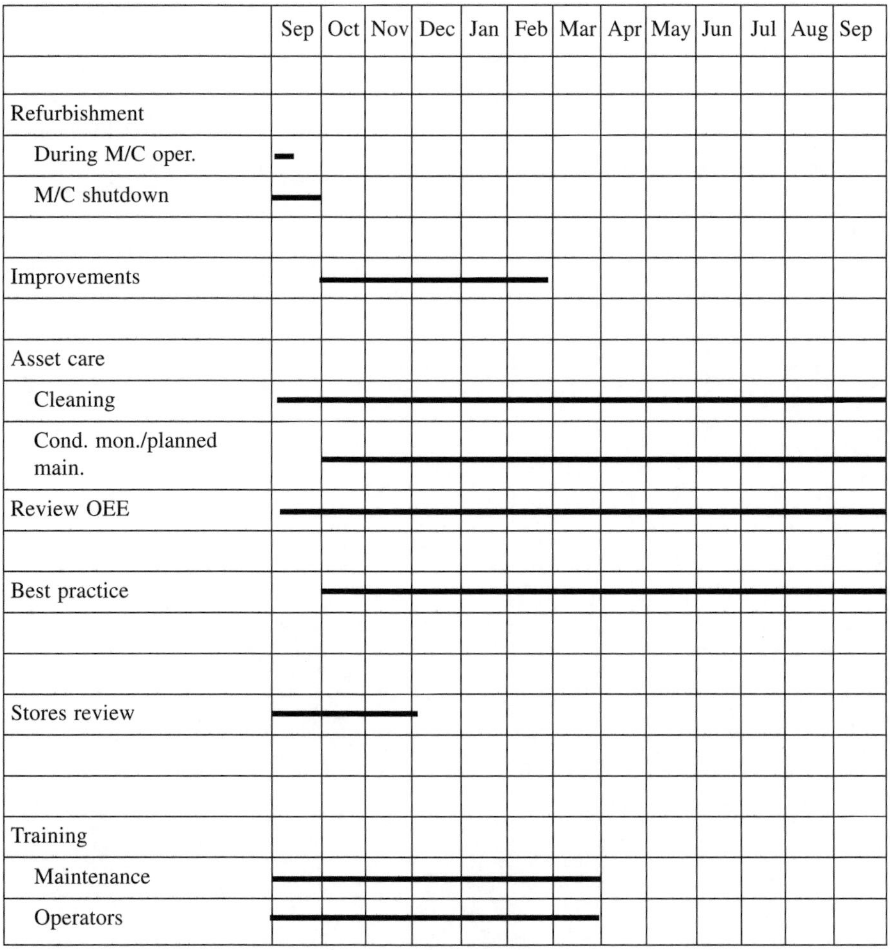

	Sep	Oct	Nov	Dec	Jan	Feb	Mar	Apr	May	Jun	Jul	Aug	Sep
Refurbishment													
During M/C oper.													
M/C shutdown													
Improvements													
Asset care													
Cleaning													
Cond. mon./planned main.													
Review OEE													
Best practice													
Stores review													
Training													
Maintenance													
Operators													

Figure 6.23 Miller M201 implementation programme

6.15 Conclusions

To achieve world-class performance, the company must move forward to:

- achieve a total quality process;
- satisfy customer requirements;
- consider all internal and external factors;
- meet internal customers and suppliers and agree requirements.

TPM is the vehicle to deliver customer satisfaction and to secure the company's future and jobs. In this pilot project, the meaning of TPM may be said to have moved from 'Today's Problematic Miller' to 'Tomorrow's Perfect Miller'!

Suffice to say that this syndicate team won the award of 'best presentation'. adjudged by their other three syndicate team colleagues on the final day of their four-day TPM workshop.

7

Planning and launching the TPM pilot

7.1 Overview

To introduce the TPM principles, philosophy and practicalities into an organization, a structured, common-sense, step-by-step approach has to be taken. This is called the *TPM Implementation Process*.

This chapter sets out an overview of the TPM implementation process. The schematic shown in Figure 7.1 should be read in connection with this chapter as we describe the detail of each building block and tools.

As mentioned earlier, it is important that the TPM implementation process builds on current good practices and to do this in a way which develops:

- *ownership* of the need for continued change and business performance improvement;
- *commitment* to use TPM as a key part of the change process;
- *skills and capability* in applying TPM techniques.

The pilot process is designed to fulfil the above objectives by creating a model of what TPM can achieve. This also supports the development of

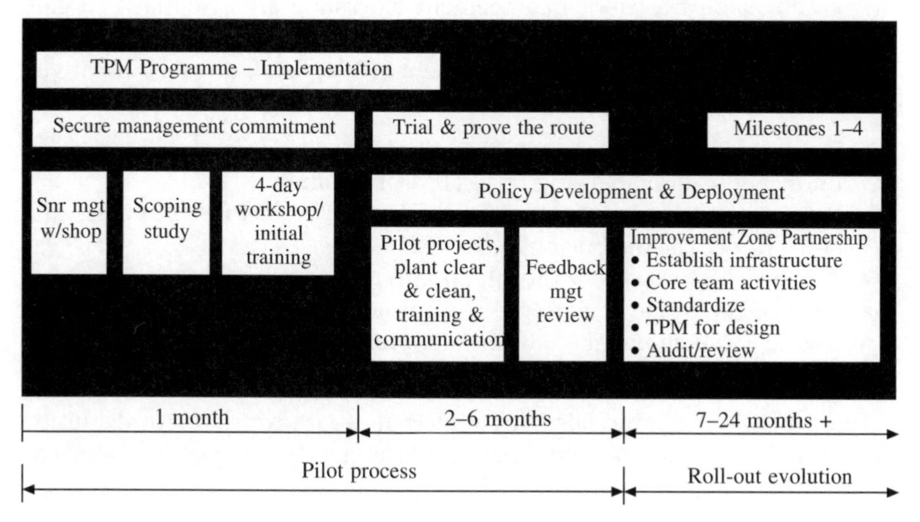

Figure 7.1 TPM implementation process

capability and experience to ensure that the roll-out process is both realistic in terms of priority, pace and resources and achievable in terms of objectives.

The pilot selection process and the CAN DO/5S improvement zone activity will also ensure that the implementation of TPM is customized and builds on existing progress and good practices and the 'logic' of the organization structure.

Chapter 3 presented the implementation process as a journey which comprises three main elements (as shown in Figure 7.1 for convenience):

- Securing management commitment
- Trialling and proving the TPM route as part of the policy development
- Deployment of that policy through four milestones, based on the geographic improvement zones

The implementation timescales shown in Figure 7.1 will of course vary according to the size of the operation, the amount of resources committed and the pace at which change can be initiated and absorbed. All these key questions, plus cost/benefit potential, are addressed within the scoping study or 'planning the plan' phase.

Thorough planning is an essential forerunner for successful implementation. Likewise, if you do not secure senior management commitment from the outset, then do not start the programme. The raising of expectations and the likely high risk of failure without that commitment are the issues at stake here.

7.2 Securing management commitment

Senior management workshop

The objectives of the senior management workshop are essentially to gain senior management commitment to the TPM process and how TPM will help deliver the business drivers and complement or integrate with other initiatives. In more detail, this means the workshop should address the following elements:

- Familiarize senior management with the principles of TPM and the implications of embarking on a TPM programme.
- Review current plans and initiatives and how TPM fits into these and helps to deliver the business drivers.
- Agree a TPM vision for the site/plant/company.
- Set a policy framework to guide improvement and implementation.
- Define a management control system for the programme.
- Define terms of reference for the scoping study.

The workshop should also involve (where appropriate) trade union representatives so that the intentions of working as 'Partners in Change' can be seen to be happening.

Scoping study

Figure 7.2 shows the *scoping study process* which is the critical planning stage to ensure that the TPM implementation programme is adapted to suit the local plant-specific needs.

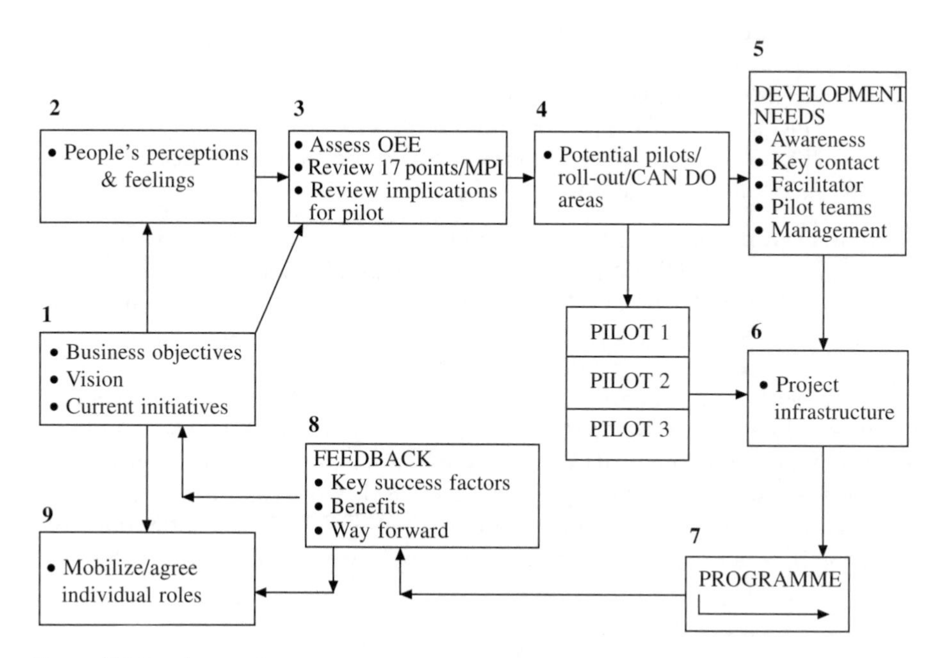

Figure 7.2 Scoping study process

The principles and philosophy of TPM are well proven. The key is to tailor the TPM suit of clothes to fit the body or plant. If you force your body into the 'off the peg' TPM suit of clothes, it will probably not fit at all well. The tailoring of the suit, however, is not corrupted so much that it becomes unrecognizable as a suit of clothes: the founding principles and pillars of TPM are still valid.

Another way of considering the scoping study is to think of TPM as a human heart: in a successful heart transplant, if you don't match it to the patient, you get rejection!

The objectives of the plant-specific scoping study are to:

- set down precisely how TPM will help achieve the business drivers and fit with other initiatives;
- assess equipment losses for potential improvement;
- carry out a cost/benefit appraisal;
- assess people perceptions and readiness for the TPM programme;
- identify TPM pilot opportunities and priorities;
- identify the critical success factors and how TPM will fit;

- develop a site roll-out approach;
- develop implementation and training plans to cover:
  - potential pilot(s)
  - likely benefits from pilots (plus potential for site-wide TPM)
  - team size and membership
  - key contact membership and roles
  - logistics and resources plus costs
  - initial awareness, communication and training plan plus timing
  - TPM facilitator support requirements plus training
  - TPM Steering Group membership and terms of reference
  - TPM pillar champions

As an example of Step 1 in the scoping study process, Figure 7.3 shows the potential impact of TPM on this company's vision of the year 2002.

Figure 7.4 articulates how TPM will provide the mortar between the other building blocks already in place at this company as their 'site improvement strategy for sustainable growth'.

The second stage in Figure 7.2 is to assess people's perceptions and feelings. What people think is what matters. Whether we agree with their perceptions is neither here nor there. If somebody has a so-called 'negative attribute', then it is our belief that they have a perfect God-given right to feel that way, because something or someone has made them feel that way. Our collective responsibility is to turn that negative energy into something positive.

VISION FOR 2002	Potential impact of TPM
◆ **World-class performance:**	
– Satisfied customers, plus competitive costs	4
◆ **Teamworking culture:**	
– Employees contribute to and share success	4
– Flexible but not dependent on overtime	3
– Few layers and better communications	3
◆ **Best use of resources by investing in:**	
– People	4
– 21st century equipment	3
– 21st century IT	2
◆ **Plan for growth:**	
– Europe	3
– New high-value productivity	3
– New business/markets outside Europe	3

1 = Little or no impact 2 = Some impact 3 = Significant impact 4 = High impact

Figure 7.3 Setting the vision

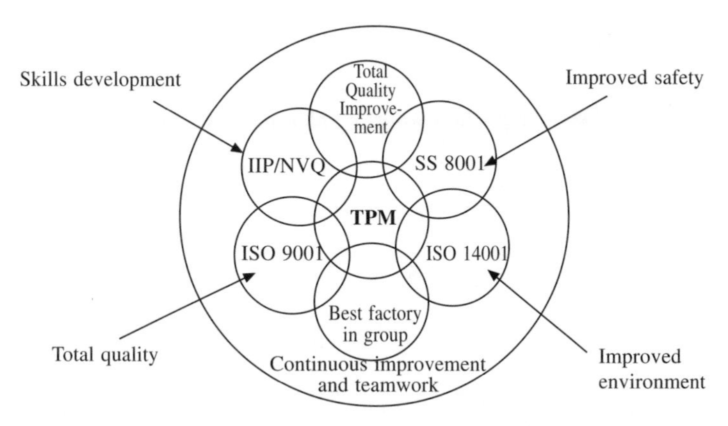

Figure 7.4 Site improvement strategy for sustainable growth

We need to get a measure of individual and collective strength of feelings and to pinpoint things that will hinder change or progress as well as the things that will positively help. The assessment takes the form of in-depth interviews and discussions to ascertain existing attitudes and, in due course, to influence those attitudes.

WCS has developed a 28-statement format in which employees (whether they are managing directors, design engineers, operators or maintainers) are asked to rank the statements from their own perspective and perception. For example, the statement might say:

'From my viewpoint, Production and Maintenance operate as separate empires.' Do you think this statement is very true, partially true or false?

Fourteen of the statements measure the employee's perception with regard to the degree of management encouragement in the organization or plant, and the other fourteen statements measure the degree of workforce involvement (see Figures 7.5 and 7.6).

It is far better to carry out these perception interviews on a one-to-one basis rather than simply giving out the questionnaire to be completed by the employee, since it gives the interviewer a chance to explain the TPM and how and when it might affect the employee. Also, it allows the interviewer the opportunity to ask a supplementary question to each of the 28 statements, such as 'Why do you feel so strongly about this statement?'. The response will often give some key directions and insights as well as, perhaps, an improving or occasionally worsening perception over time.

The subsequent analysis clearly shows the differences in strength of feeling between, say, members of management compared to key contacts and, of course, comparisons with operators and maintainers. Similarly, the strength of response across the spectrum of employees will show up quite clearly for

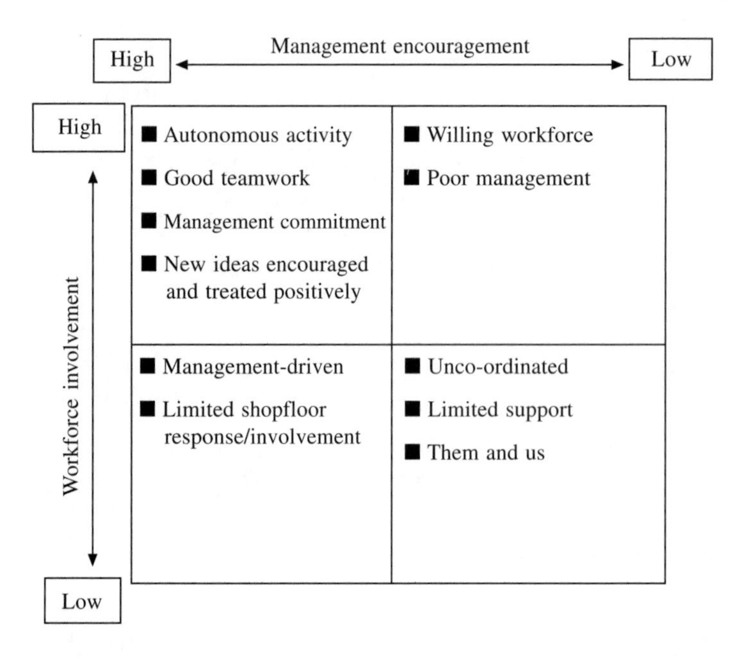

Figure 7.5 Analysis of 28 statements

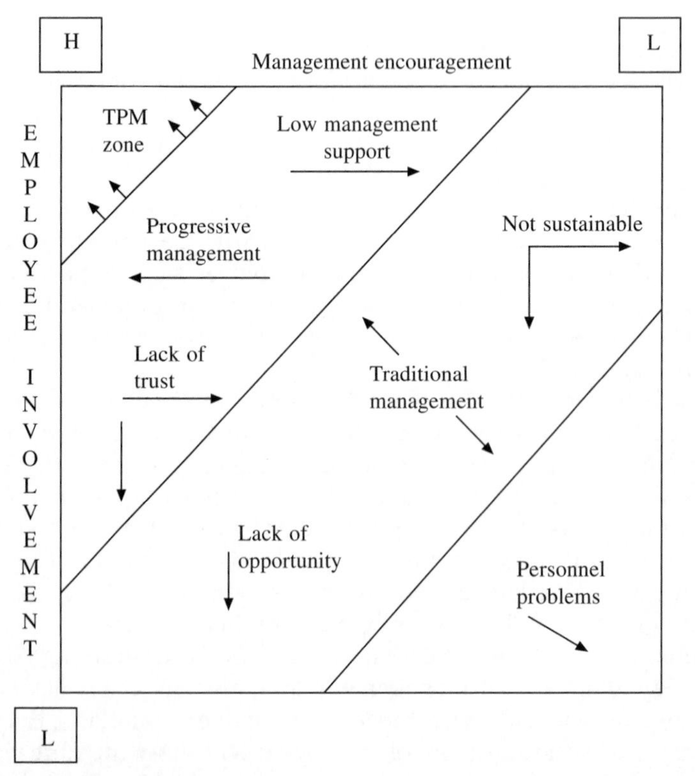

Figure 7.6 Perceptions matrix implications

each of the 28 statements. These can then be grouped as 'hopes' and 'fears' as well as potential TPM 'hinders' and 'helps'.

The analysis will position employee groups on the matrix; the higher the grouping towards the left-hand corner, the better for TPM's likely acceptance and success. However, if the groupings are towards the lower end of the horizontal and vertical axes, you will find that the TPM process addresses many of the perceived hindrances in a positive and lasting way. This perceptions tool is not absolute, but it does provide an excellent benchmark against which to measure future movements on the matrix.

The analysis will have a major bearing on the way the TPM process is implemented. As the plan develops, the training programme will seek to ensure that the most constructive and progressive attitudes prevail, firstly in the pilot project (see later) and then company-wide as the TPM process develops.

Achieving the right attitude to change is essential for success. Experience has shown that operators, recently engaged staff and younger people tend to take a positive attitude to change, whereas the old hands and the experienced maintenance technicians are likely to be more wary and defensive (Figure 7.7). The attitude of supervisors depends very much on the individual. Supervisors will normally support the idea of TPM because of its common sense. However, they have to face the day-to-day demands of production and quality and, hence, may find it difficult to sustain a commitment to release operators and maintainers for the TPM process or to release equipment and machines for essential restoration and refurbishment. Effective two-way communication is essential to avoid resistance to change: those who will be involved in the TPM process must have a very clear idea of what it is all about and what the company – and, more particularly, what they as individuals – stand to gain. Resistance must be broken down by explanation, thorough discussion and the establishment of total confidence in the eventual outcome (Figure 7.8).

Figure 7.9 depicts the way in which resistance can be broken down by ensuring the full involvement of the people concerned and by securing their enthusiasm and dedication. Effective communication is more than the one-way approach of informing people and preparing them for change.

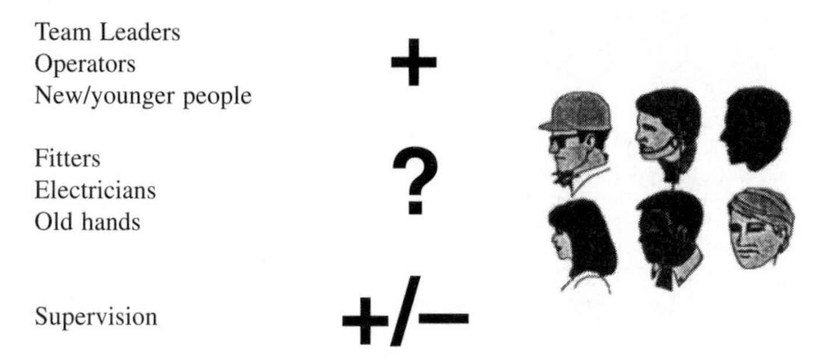

Figure 7.7 How they see TPM

POOR COMMUNICATIONS = RESISTANCE

- The less I know about plans to change
- The more I assume
- The more I assume
- The more suspicious I become
- And the more I direct my energy into

RESISTANCE

Figure 7.8 Causes of resistance

People need to be part of the change process, actively involved in decisions and able to influence the outcome. If treated with respect, they gain recognition and self-esteem and a two-way communication will result. They may well need to get the acid or bile which has accumulated over the years out of their system, but essentially employees *want* to improve their lot. A typical reaction came from a team leader who, after three months' involvement in the TPM process, commented that TPM is built on teamwork and that 'Today People Matter'.

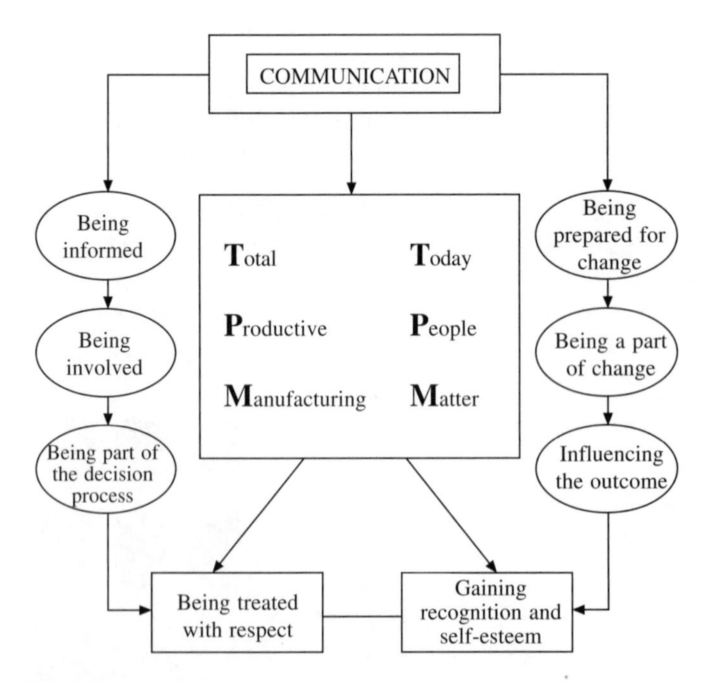

Figure 7.9 Effective communication

If Jack is said to have an attitude problem, it is normally assumed that he has a negative attitude. He has every right to feel that way and may well have been influenced by developments over a number of years. It will take time to change negative energy into positive energy, and TPM may well be the catalyst to move Jack from left to right in Figure 7.10.

Perhaps the key to influencing change and people's behaviour is to put yourself in the other person's shoes. Figure 7.11 attempts to illustrate this point by suggesting:

> I know me very well and I know you quite well. However, I do not really know what you think of me, and I have no idea what you think of yourself!

NEGATIVE	POSITIVE
PHASE 1 AWARENESS	
Not interested	Seeking out new ideas
PHASE 2 COMMITMENT AND PREPARATION	
Cynical	
It won't work	Active learning
PHASE 3 TRANSFER TO REAL LIFE	
It's flavour of the month	Objective testing
PHASE 4 TAKE-OFF	
Highlighting failures	
	Making it work
PHASE 5 CONTINUOUS IMPROVEMENT	
It's only what we have always done	Refining, improving

Figure 7.10 Attitude to change as an indicator to progress

AN AID TO UNDERSTANDING MOTIVATION

	ME	YOU
How I see	✓✓	✓
How you see	?	??

The gift . . .
. . . to see ourselves as others see us

Figure 7.11 Developing a full picture

The gift is to see ourselves as others see us.

Figure 7.12 shows the spread of perceptions from an actual scoping study and highlights the fact that:

- 70 per cent feel encouraged and involved;
- only 8 per cent feel neither encouraged nor involved;
- TPM will directly address six of the top seven perceived 'hinders'.

Figure 7.13 shows the top seven perceived hinders to change and Figure 7.14 highlights the top seven helps towards combined progress.

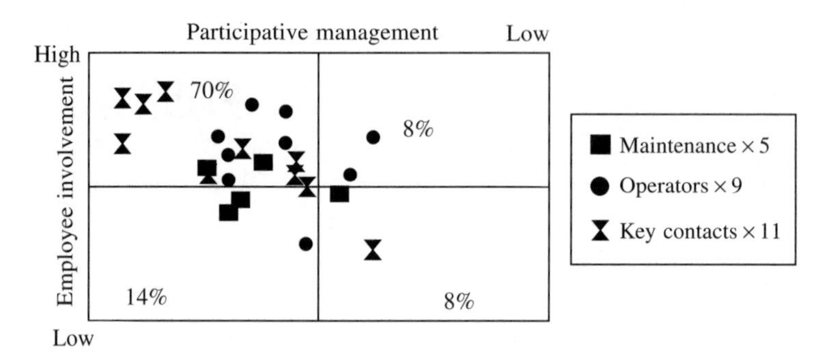

Figure 7.12 People's perceptions and feelings

	Maintainers	Operators	Key contacts	Total
7 People are reluctant to say what they really think	80%	81%	55%	69%
10 Skills are picked up rather than learned systematically	40%	85%	64%	67%
18 Inter-departmental communication is poor	100%	63%	48%	64%
1 Production and maintenance are separate 'empires'	93%	41%	58%	59%
31 We suffer from too many initiatives	60%	78%	42%	59%
21 Lessons learned on one shift do not get transferred to others	87%	37%	61%	57%
33 Work organization problems are not faced openly and frankly	40%	56%	55%	52%

Figure 7.13 What hinders progress?

	Maintainers	Operators	Key contacts	Total
40 We should introduce a TPM approach	100%	89%	100%	96%
48 Standard methods are seen as important	100%	85%	97%	93%
45 The company takes safety seriously	87%	85%	94%	89%
47 The company tries to make jobs interesting	100%	93%	76%	87%
32 People welcome more challenge in their jobs	60%	78%	85%	77%
43 Unit cost information is made available to me	87%	70%	70%	73%
42 Most of my work is planned	53%	78%	76%	72%

Figure 7.14 What helps progress?

The key conclusions are that:

- most feel that 'across the fence' communications and understanding of 'each other's problems' have some way to go (see Q7, 18, 1, 21 and 33);
- most *expect* change to continue (see Q40, 47 and 32), although some may still resist it because they already feel there is too much going on (see Q31).

Figure 7.15 shows the 20 statements' strength of feeling as to how the individual feels he or she is treated and how well the team management and company works as far as he or she is concerned.

How I feel that	Maintainers	Operators	Key contacts	Total
That I am treated	45%	38%	30%	36%
How the team works	55%	38%	40%	42%
The management works	37%	40%	34%	37%
The company works	36%	44%	24%	34%
EI	50%	38%	35%	39%
PM	37%	42%	29%	35%

The lower the score the better

Figure 7.15 People's perceptions and feelings

A further key part of the scoping study (see Step 3 of Figure 7.2) is to gain an assessment of existing levels of OEE and the potential for improvement.

One of the most powerful ways of gaining a 'snapshot' OEE is to carry out a study of a machine, line or process by spending at least a whole shift on the specific plant item in order to see the reality of the actual situation.

Figures 7.16, 7.17 and 7.18 show the valuable outputs of such a study.

Figures 7.19 and 7.20 show the results of a five-month reference period analysis to set realistic challenges on downtime and scrap improvement.

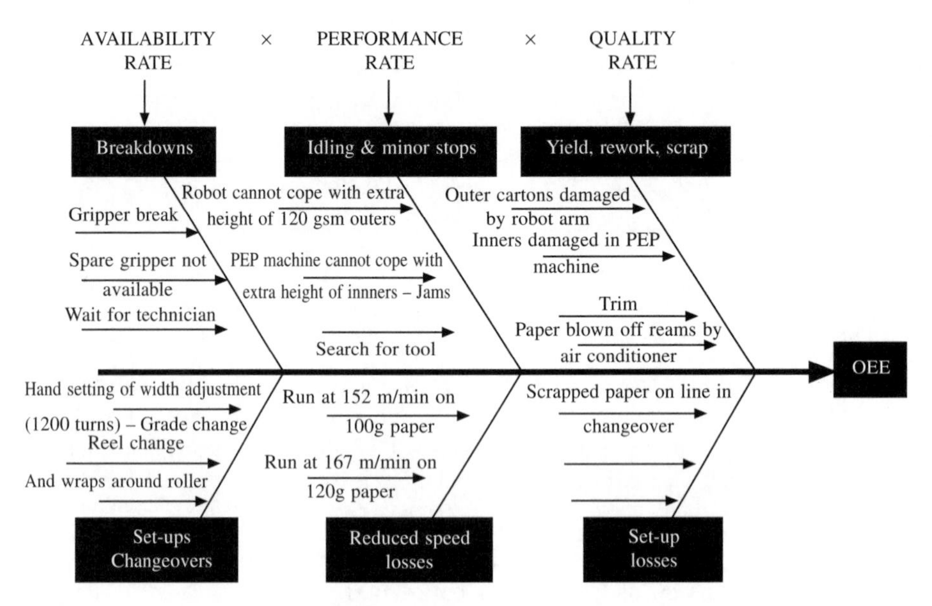

Figure 7.16 Line study: Assessing the six losses

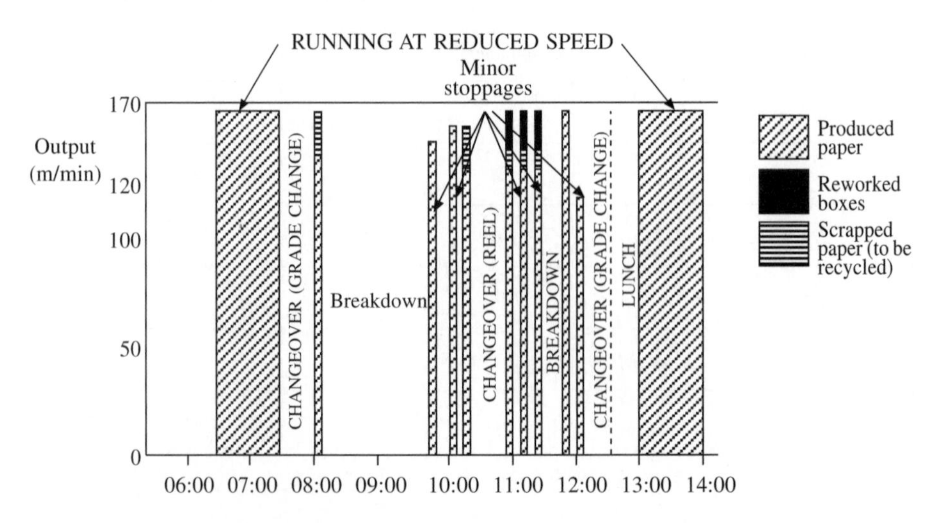

Figure 7.17 Line study summary

Av × **Perf** × **Quality** = **OEE**	
32.7% × 84.2% × 88.6% = 24.4%	
PLANNED RUN TIME 450 mins	
AVAILABLE TIME 147 mins Estimated output 1622 reams	BREAKDOWNS & CHANGEOVERS
ACTUAL OUTPUT 1366 reams	RUNNING AT REDUCED SPEED MINOR STOPS (RESETS, JAMS)
RIGHT FIRST TIME OUTPUT 1210 reams	REWORKED/RECYCLE/ SCRAP PAPER

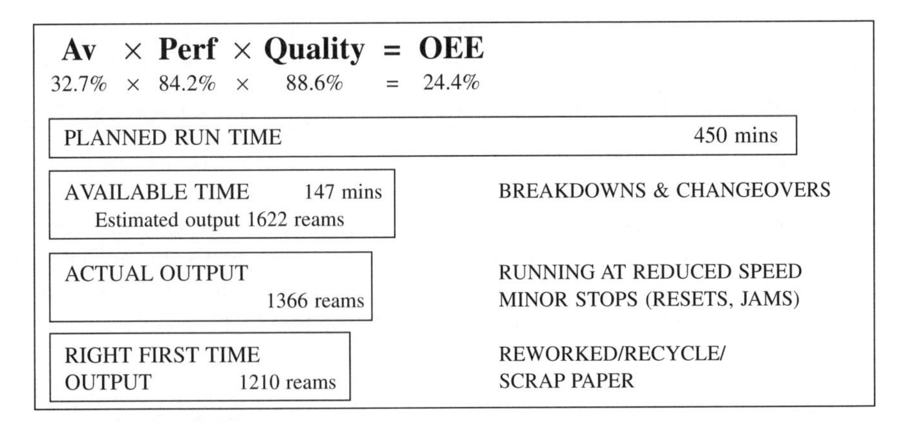

Figure 7.18 Line study OEE calculations

- 233 interventions/downtime events (11 per week)
- Of average duration 1.4 hours every 13 hours
- 59% less than 1 hour duration
- 25% less than 30 mins duration

THE CHALLENGE

By end of 1999: MTBF = 20 hours MTTR < I hour
Interventions less than 5 per week

THE RESULT DOWNTIME @ 11% ➡ 5%

Figure 7.19 Downtime analysis: 1 January – 31 May

THE CHALLENGE: By end of 1999, at the rate of:

A Reduce our inside controllable losses by 25%
B Reduce trim by 10% of current levels
C Reduce our external (supply) losses by 10%

THE RESULT

Will save annual rate of:

A	45 tonnes
B	20 tonnes
C	10 tonnes
	75 tonnes

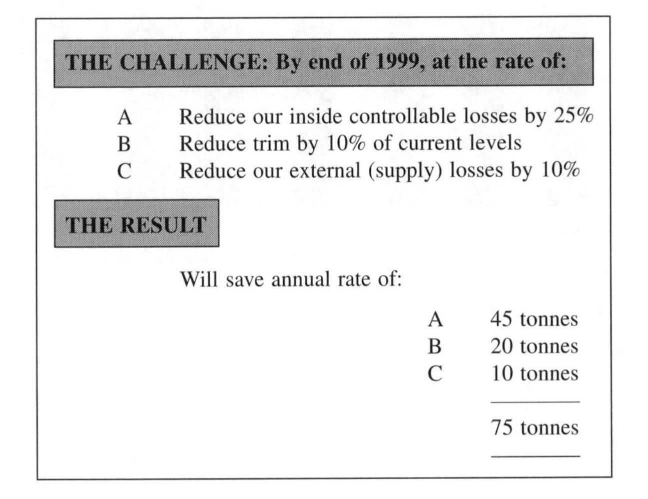

Figure 7.20 Scrap analysis

Figures 7.21 and 7.22 show how the pilot project will be organized over four shifts to tackle this production line using the nine-step TPM process over sixteen weeks as a pilot project.

The key point to stress is the commitment of time for each shift over sixteen weeks dedicated to the TPM process – equivalent to 10 per cent of attendance time to the process as a key learning experience.

It is a fact of life that you will not make sustainable improvements in anything you do unless you dedicate improvement time to the activity: it cannot be done 'on the run'. If management will not commit to this reality, then you should seriously consider not starting the TPM process, as it is likely to fail.

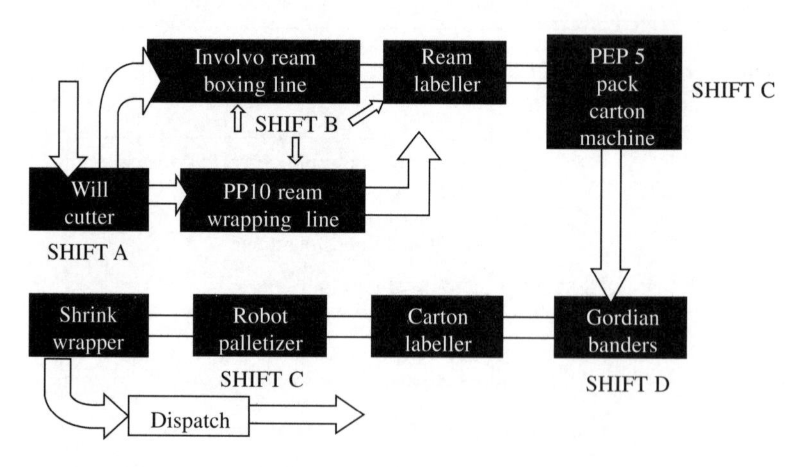

Figure 7.21 Line process layout: Proposed pilots

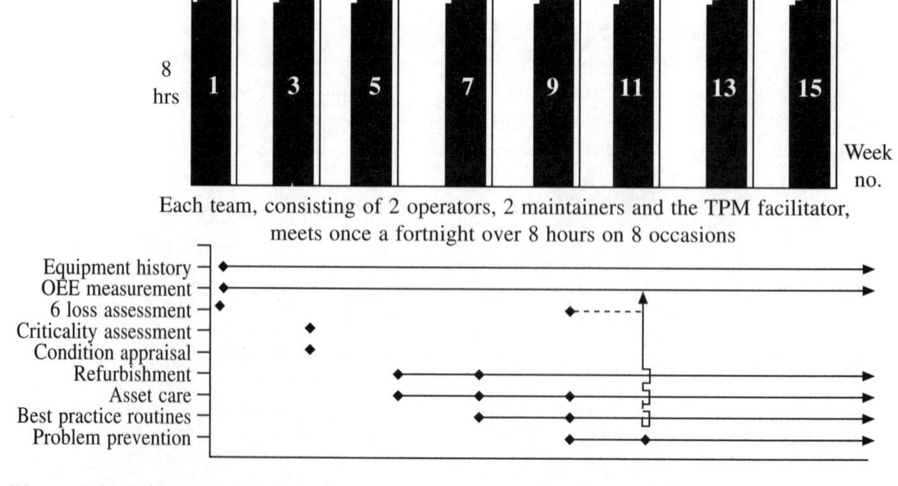

Figure 7.22 Pilot project structure

Four-day hands-on workshop

The final stage of securing the necessary commitment is a four-day hands-on TPM workshop carried out on 'live' equipment in the host plant. The delegates will comprise a cross-section of senior management, potential TPM facilitators, union representatives (if appropriate) and some key operators and maintainers. The content of the workshop is shown in Figure 7.23.

The objectives of the four-day 'in-house' workshop are:

- to provide a thorough understanding of TPM and how to put it into practice;
- to provide a framework and understanding for TPM facilitators to work in, and influence the behaviour of, multi-discipline, multi-interest teams;

DAY 1	DAY 2	DAY 3	DAY 4
• Introduction to TPM Ex – Whose Job? • TPM PRINCIPLES • Slide Presentation Ex – OEE Ex – Project Management	• Recap Brief Visit to Pilot and Plan the Plan Equipment Description • MEASUREMENT CYCLE (On-the-Job) ♦ History/Records ♦ OEE Measures ♦ 6 Loss Assessment	• Building TPM Activity Boards • Consolidatiog Measurement and Condition Cycles • SUPPORTING TECHNIQUES ♦ 5 Whys ♦ P–M Analysis ♦ Visual Management ♦ Set-up Reduction	• Dry Run Presentations • SYNDICATE PRESENTATIONS • REVIEW AND KEY LEARNING POINTS • Ex – Whose Job?
• TPM TECHNIQUES AND 9-STEP IMPROVEMENT PLAN • 5S/CAN DO • CASE STUDY • TEAMWORKING & FACILITATING Ex – Active Listening • BRIEFING FOR SYNDICATES On-the-Job Case Study	• CONDITION CYCLE Ex – Criticality Assessment (On-the-Job) ♦ Criticality Assessment ♦ Condition Appraisal ♦ Refurbishment Plan ♦ Asset Care	• PROBLEM PREVENTION CYCLE (On-the-Job) ♦ Best Practice Routines ♦ Problem Solving • PREPARE PRESENTATION	• PILOTS AND 4-STAGE ROLL-OUT • SUPPORTING THE PILOTS • Course Assessment • 16.00 hrs CLOSE

Figure 7.23 TPM four-day hands-on workshop programme

Managing Director Manufacturing Director Business Unit Manager × 3 Design Manager Engineering Manager Mechanical Engineer Line Operators/Maintainers × 8
TOTAL NUMBER OF ATTENDEES = 16

Figure 7.24 Attendees at four-day workshop

- at least 70 per cent of the workshops are focused on carrying out the TPM process on your own 'live' TPM pilot equipment on-site so that delegates can experience the on-the-job reality of putting TPM into practice.

Sales & Marketing Manager
Technical Dept Representative
Development Manager
Client Services Manager
Commercial Manager
Finance Manager
Reproduction Manager
Computer Systems Development Supervisor
Design & Improvement Supervisor
CAD Designer
Technical Dept Supervisor
Data & IT Supervisor

Figure 7.25 Attendees at key contact training session

Finally, at Figures 7.26 and 7.27, we show an example of the Steering Group membership and their terms of reference.

Managing Director
Manufacturing Director
Business Unit Manager
Design Manager
Engineering Manager
Site Union Convenor

Figure 7.26 TPM Steering Group

All key points of the scoping study process, including the initial programme, are shown in Figure 7.28. This shows a typical site-level bar chart schedule of activities over the first six months of the TPM programme – namely to secure management commitment and to trial and prove the TPM process through the pilot projects, as a key learning experience before launching the roll-out programme.

During the pilot projects, the TPM policy for roll-out will be developed as part of the pillar champions' input into the TPM master plan for future policy deployment through the four milestones. This is detailed in Chapter 8.

7.3 Trial and prove the route (refer again to Figure 7.1)

Following management's acceptance of the scoping study plan, the pilot

Given below are 'Terms of Reference' for the XYZ Company TPM Steering Group. The Group should comprise no more than six members.

- To provide the Executive with a regular TPM Progress update
- To communicate and co-ordinate the TPM Activity at the XYZ Company
- To establish the level of resource and infrastructure to support the TPM Activity and agree the key success factors
- To set up and monitor the necessary TPM and CAN DO Audit & Review and Measurement Process
- To establish common TPM and CAN DO Training Programmes, Materials, Procedure and Completion Criteria (from Pilots → Roll Out → Continuous Improvement)
- To ensure a regular and detailed update of the TPM Pilot Projects and CAN DO Improvement Zone Progress
- To ensure that any roadblocks to Progress are addressed, action agreed and resolved
- To confirm TPM Pilot Meeting dates are set and to ensure that these meetings take place and that equipment is released for any essential planned refurbishment
- To establish the TPM Policy & Procedure for the XYZ Company operations

Figure 7.27 Steering Group terms of reference

Schedule of activity to secure management commitment and trial and prove the route

	Secure management commitment				Trial and prove the route				
Pilot process	Week 1	Week 2	Week 3	Week 4	Mth 2	Mth 3	Mth 4	Mth 5	Mth 6
Snr management workshop	↔								
Scoping study		←	→						
Management review				↔					
Awareness & communication					↔				
Training – 4-day workshop & key contacts					←	→			
Launch pilot projects					←	—	—	—	→
Plant clear and clean/5S/CAN DO					←	—	—	—	→
Project management & coaching					←	—	—	—	→
Formulation of roll-out							←	—	→
Management reviews & approval of roll-out									↔

Figure 7.28 Typical TPM timetable plan

projects and associated CAN DO improvement zones will be mobilized, together with the supporting activities of:

- awareness and training, including an immediate on-site four-day hands-on workshop for facilitators, key contacts and members of the Steering Group, together with the core team nominees;
- policy and roll-out plan development;
- on-the-job coaching;
- Steering Group/Audit review.

The pilot projects themselves support a number of implementation processes:

- Training for the core teams
- On-the-job coaching for the core teams, team leaders and facilitators
- Identification of issues which restrict the application of TPM principles
- Integration of TPM with existing internal systems and procedures
- Development of the policy and roll-out plan to support the systematic implementation of TPM across the site

This is the key phase for moving the TPM process from 'Strategic Intent' to 'Making it Happen', concentrating on focused improvements on the pilots using WCS's unique nine-step Improvement Plan and getting everyone involved via the plant clear and clean activities of the 5S/CAN DO philosophy. This phase also includes setting up the TPM infrastructure, including the Steering Group, TPM facilitator and TPM pillar champions.

The objectives of the TPM pilot training are to:

- conduct communications and awareness sessions;
- implement the selected pilots using the nine-step improvement plan, based on the three cycles of measurement, condition, problem prevention;
- design, develop and implement a plant-wide clear and clean process using 5S/CAN DO philosophy;
- establish performance and measurement to record progress, with specific audit and reviews;
- establish infrastructure to support eventual site-wide deployment of TPM, including pillar champion roles, responsibilities and TPM coaching needs;
- gain experience and identify key learning points;
- highlight the inhibitors to effective implementation for action;
- ensure that the policy guidelines defined earlier are applied;
- monitor and review progress with the Steering Group.

As shown in Table 7.1, the core teams will introduce low cost/no cost improvements throughout the pilot over twelve to sixteen weeks. During this time, they will need to meet a minimum of eight times, following our structured nine-step approach. At the end of the period, they will feed back their recommendations for future action and also their views on the effectiveness of TPM.

Table 7.1 Typical TPM pilot project timetable

Week No	*Content*	*Support activities*
1	Initial training and pilot selection equipment history	• OEE definitions
2 to 3	OEE evaluation/assessment of 6 losses	• TPM activity board
4 to 5	Criticality assessment/condition appraisal	• 5S/CAN DO clear and clean activity
6 to 7	Refurbishment plan/asset care	• Refurbishment plan • Asset care training
8 to 9	Best practice routines/problem resolution	• Refurbishment action • Trial improvements • Single-point lessons • M/C visual help/aids
10 to 11	Prepare for feedback/presentation	• Refurbishment action (cont/d) • Trial improvements (cont/d) • Training plan
12	Feedback dry run/presentation	

In parallel, and following on from the general awareness sessions, all shopfloor personnel in the identified geographic improvement zones (determined largely by the logic of the team leader's span of control) will be involved in workplace organization activities, starting with a plant clear and clean activity. This aims to reinforce the key learning points from the training in a way which raises existing housekeeping standards and introduces the concept of shopfloor ownership in a hands-on way, based on the existing shift-based geographic zone. The activity also includes a very detailed and structured audit and review of the CAN DO process to target areas for improvement and to ensure that the gains are held.

Implementation of improvement zone

The pilot provides management with the experience to identify gaps in general areas of best practice. This results in the generation of management standards which can be translated into local policy at a shopfloor improvement zone. See Table 7.2.

The improvement zone implementation progress can then be measured against these standards, providing a basis for team-based recognition at each level.

If top-down management job descriptions and personal development plans are amended to reflect success at each improvement zone level, this effectively ties in top-down and bottom-up recognition systems.

Table 7.2 TPM bottom-up standards

Step	*OEE*		*OAC*	*MAC*		*Continuous skill development*		*Early equipment management*
	Loss deployment	*Focused improvement*		*Planned maintenance*	*Quality maintenance*	*Function development*	*Safety deployment*	
1	Highlight loss levels, priorities KPIs, cross-shift accountabilities and progress reporting	Support technical problem/improvement activities in initial roll-out phases to address sources of contamination	Initial cleaning of workplace and equipment/condition appraisal	Maintenance WPO refurbishment and critical system back-up routines	Technical documentation, critical assessment. Define key checkpoints/preventive maintenance	Actions to formalize current practices across the shifts	Safety assessment to align actual and current practices (plus future needs)	Actions to identify design, technology and project management losses (EEM policy)
2	Establish team-based performance management at all levels linked to the current year business planning process	Transfer lessons across similar equipment	Action of source, including cleaning and use of SPLs	Contain accelerated deterioration, develop counter measure, including correct parameter setting	Analyse/address accelerated deterioration and impact of improve 6 LCC loss factors include ease of use, etc.	Actions to standardize core competences, including correct operation and basic maintenance techniques	Actions to reduce safety risk and promote behavioural safety as part of CI	Actions to measure LCC reduce losses. Integrate equipment management roles (FI, QM, EEM) and establish knowledge base
3	Integrate future business planning with PDP and specific loss reduction targets	Focus on support problem/improvement activities, including PDP targets	Adopt apple a day standards and use of visual indicators	Set thermometer/injection needle standards to improve response time and feedback	Correct design weaknesses to improve precision and feed back to knowledge base	Actions to simplify, combine, eliminate	Refine procedures using visual indicators to reduce risk and maintain awareness	Design actions to eliminate sources of contamination and support zero breakdowns
4	Define future loss vision linked to exceed future customer expectations	Assess supply chain losses. Define the future customer requirements	Adopt thermometer standards	Eliminate sporadic losses (breakdown analysis, condition-based replacement)	Improve technical documentation, raise understanding and ownership	Establish training plans to support normal conditions and future skill gap analysis	Define future safety and environmental needs. Establish normal conditions	Define the future company response. Early product management

8
Managing the TPM journey

8.1 Future vision, planning and control

The introduction of TPM to an enterprise starts with a vision of the future, and this is illustrated in very clear terms by Figure 8.1. All the means of achieving TPM which have been discussed in earlier chapters lead to the continuous improvement habit, which embodies the spirit of *kaizen* and which can be brought to reality by following the WCS approach to TPM. The key point is that when people *want* to change the way they do things, then they will sustain it.

Some of the major changes which will result from the introduction of TPM, and the benefits which those changes will bring, are as shown in Table 8.1.

Planning, organization and control are essential prerequisites:

- *Planning* entails allocation of resources on a realistic and achievable basis with regular review and progressive development on the long-term basis necessary for success.
- *Organization* requires defined resources with clear allocation of roles and responsibilities; this must be accompanied by effective and clearly understood methods of working.

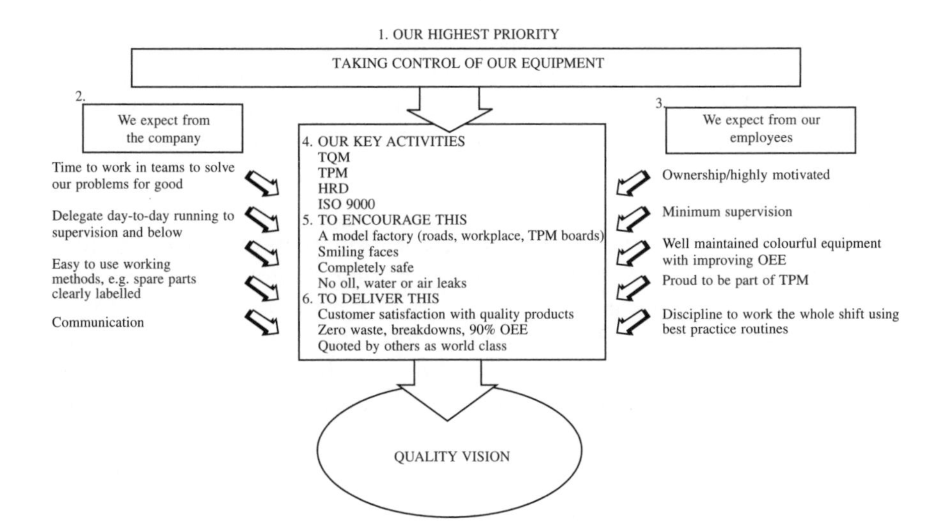

Figure 8.1 Our 'Spark to Start vision' win/win contract

Table 8.1 TPM enablers and results

Enabler	*Result*
Machines run close to name-plate capacity	Reduce need for excess capacity
Ideas to improve often proposed by operators	Ownership/success
Breakdown rate reduced	Used to learn and teach the team
Machines adapted to our needs by our people	Our machines will be better
Operators solve problems themselves	Fewer delays and stoppages
Cleanliness and pride in continuous improvement	Good working environment
More output from existing plant	More profits

- *Control* The two aspects of control are *coordination*, which is concerned with what happens next and is most effective with simple vision systems and procedures; and *feedback*, which is
  - concerned with goals for time, cost and quality
  - used to identify the reasons for failure and to prevent recurrence
  - the source of objective evidence of the need for increased resources, modification of goals or the introduction of specialists.

8.2 Role of managers

The implementation of TPM has three dimensions:

- *Top-down:* creating the environment for continuous improvement
- *Bottom-up:* small group activity
- *Organizational learning:* capturing and sharing lessons learned

These align with first line and senior management roles and provide the basis for integrating management priorities through an infrastructure illustrated in Figure 8.2. This is also aimed at giving the bottom-up, team-based activity the necessary recognition at each level of TPM progress (see Figure 8.3).

Top down

Pillar champions focus on co-ordinating the implementation of individual TPM principles by setting policy and supporting its application. Policy is about problem solving and sets out a fluid set of ground rules in the form of priorities and standards. TPM provides the tools to deploy that policy, translating top-level perspective stepwise into shopfloor accountabilities through the first line management or area champions. An outline of these standards is included in Chapter 7.

Bottom-up

First line managers are allocated physical areas in which to focus their improvement resources. Their role is to develop the capability of multi-

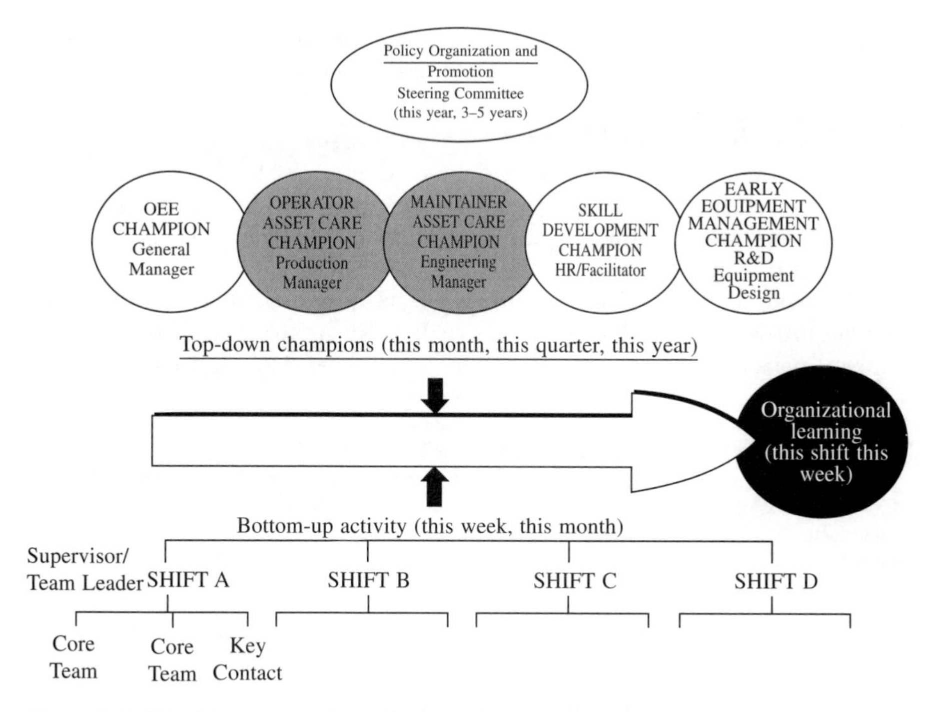

Figure 8.2 TPM infrastructure/roles for a continuous improvement habit

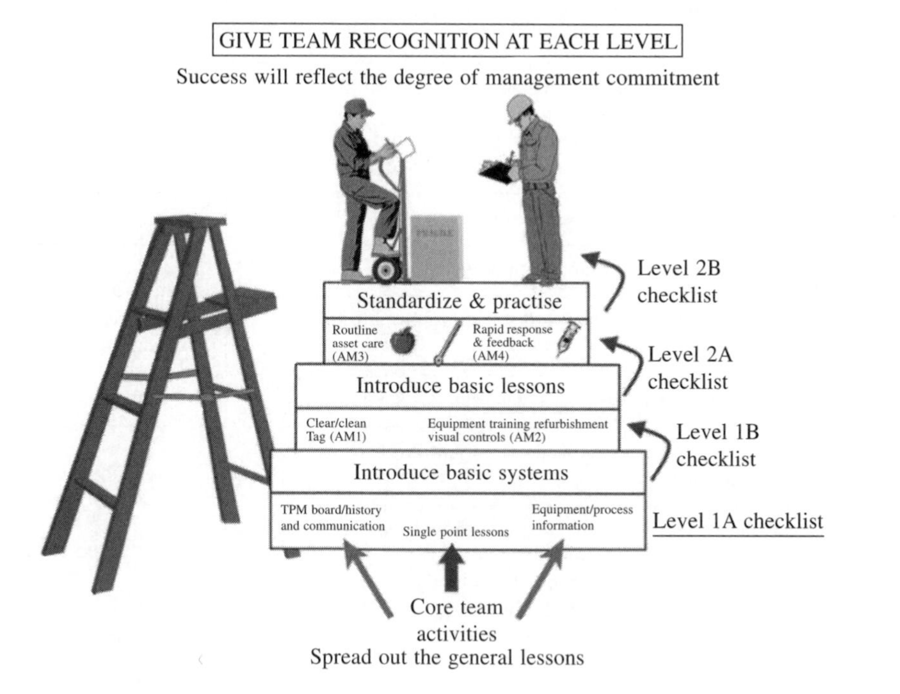

Figure 8.3 Give team recognition at each level

discipline teams of five to seven personnel. These teams will direct a minimum of 5 per cent of their time to continuous improvement.

Organizational learning

Often first line management is perceived as the barrier to change. In reality, 'what gets measured gets attention'. Traditionally first line management is left alone provided the tonnes go out of the door. Anything else is a 'nice to have', and if it doesn't happen, then it will be ignored.

TPM overcomes this by measuring progress against quality milestones (see Figure 8.3) based on evidence of bottom-up progress through the improvement zone implementation steps (see Figure 8.4). Figures 8.4 and 8.5

TPM IMPROVEMENT ZONE AUDIT/REVIEW

Milestone: Planning/Mobilization — Level:
Department: — Zone:
Auditors: — Date:

	Review point	Assessment		Evidence
		Yes	No	
1	Are improvement areas and zones clearly defined?			
2	Are zone production capacities/bottlenecks identified?			
3	Have weaknesses in documentation been assessed?			
4	Has an assessment of improvement zone benefits been made/documented and priorities defined?			
5	Has future TPM vision been clarified?			
6	Are team leaders allocated to improvement zones?			
7	Have facilitation responsibilities and resources been identified?			
8	Has a firm timetable of activities been developed?			
9	Has an assessment been made of current levels of housekeeping?			
10	Has the TPM information centre been updated?			
11	Has a roll-out cascade been defined by the team leader for each improvement zone?			
12	Have teams been briefed?			

Figure 8.4 Mobilization checklist

TPM IMPROVEMENT ZONE AUDIT/REVIEW

Milestone: Introduction | Level: 1A
Department: | Zone:
Auditors: | Date:

	Review point	Assessment		Evidence
		Score	No	
1	Is a TPM board in place for each improvement zone?		5	3 = up to date 4 = improved
2	Safety procedures defined		5	3 = used 5 = improved
3	Workplace initial clean (CAN DO Step 1)		5	CAN DO audit results
4	Equipment initial clean		5	5 = maintained
5	Cross-shift supervisor prioritization		5	3 = agreed
6	Identification of frequent problems (6 losses) and root causes		5	3 = recorded 5 = improvement
7	PLC/computer software back-up		5	3 = available
8	Equipment description (sketch, critical areas, parameters, process flow chart)		5	3 = acceptable 5 = understood
9	Checkpoints (e.g. pressure, temperature, RMP) and preventive maintenance schedule		5	3 = available
10	Problem register in place recording equipment history, including identification of accelerated deterioration		5	3 = recording up to date 5 = reduction in stoppages
	TOTAL		**50**	

Minimum score 30 = level 1A, 40 = level 1B

Rating based on procedures/systems which are:

1 Not in place, with no plans to address

2 Weak/deficient

3 Able to meet departmental/plant goals with plans to improve

4 Well defined, executed and understood

5 Well defined, with a track record of continuous improvement

Figure 8.5 First-level bottom-up audit criteria

are sample checklists to support the launching and initial auditing of an improvement zone. As this requires the active co-operation of management the rate of progress is a measure of the degree of alignment between top-down and bottom-up priorities. The rate of progress is, therefore, also a measure of organizational learning (see Figure 8.6).

The management role can be summarized as three activities, as shown in Figure 8.7.

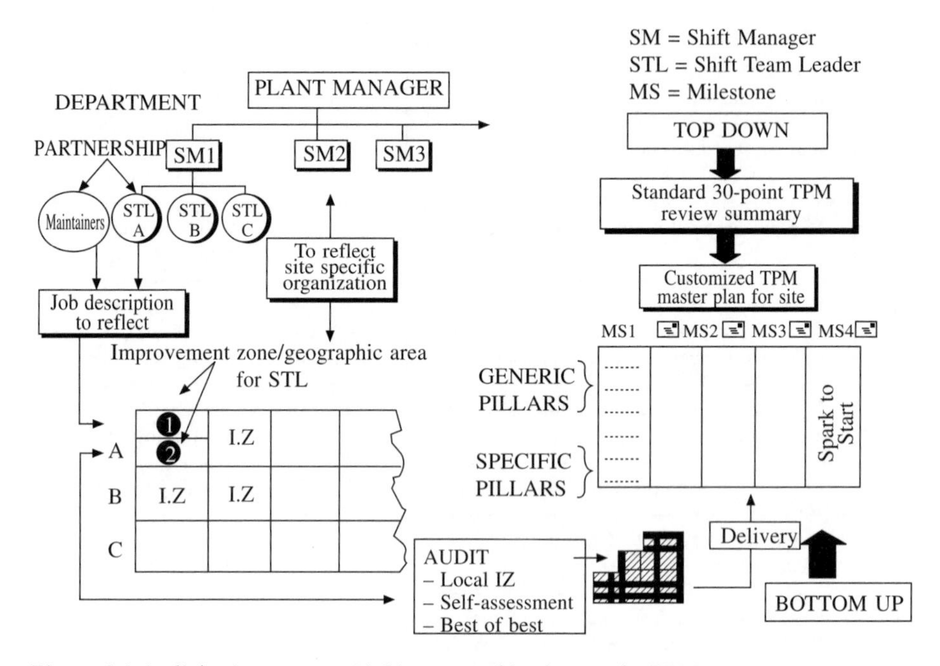

Figure 8.6 Audit/review process: Linking team objectives to the TPM

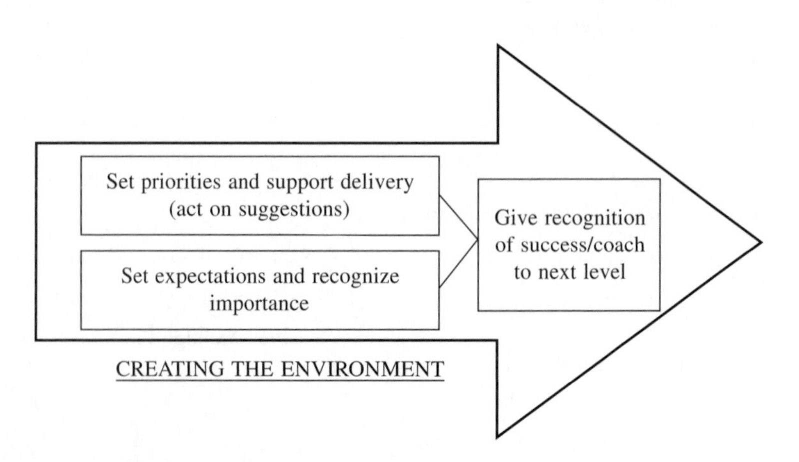

Figure 8.7 Top-down champion role/process

These priorities are linked to the future business vision through the use of a continuous improvement master plan. The master plan is simply a summary of intentions laid out against the predictable stages of the TPM change programme. These are shown in Table 8.2.

The master plan also integrates pillar champion activities to deliver a single agenda for change. Each milestone of the plan provides a quality check that the management team are pulling together. Progress towards each milestone is monitored and supported by the quarterly top-down audit coaching process. This looks for evidence of progress bottom-up to highlight where top-down policy is effective or needs support (see Table 8.2).

Table 8.2 Basic structure of the TPM master plan

Milestone	*Theme*	*Activity*	*Benefit*	*Timescale*
1	Introduction	Get everyone involved	Improved ownership OEE +10 to 15%	1–2 yrs
2	Refine best practice	Standardize and simplify routine activities	Reduced sporadic loss OEE +20 to 35%	2–3 yrs
3	Build capability	Redeploy expertise to achieve Milestone 4	Increased plant capability with less intervention	3–4 yrs
4	Strive for zero breakdowns	Optimize progress	Better than new performance OEE+50 to 60%	4–5 yrs

8.3 TPM cost/benefit analysis

The impact of equipment losses ripples through the organization, touching every function and promoting reactive, inward-looking systems and processes. As equipment becomes more reliable through the application of TPM, these ways of working will not be automatically revised to reflect that fact. As Figure 8.8 illustrates, there is little merit in getting a machine OEE up from 65 per cent to 90 per cent, if the door-to-door losses stay at 55 per cent.

To address this issue, company-wide TPM considers company-wide losses under four main headings:

- Equipment
- Transformation
- Material
- Management

Equipment losses

This covers the traditional six classic losses plus design losses of operability

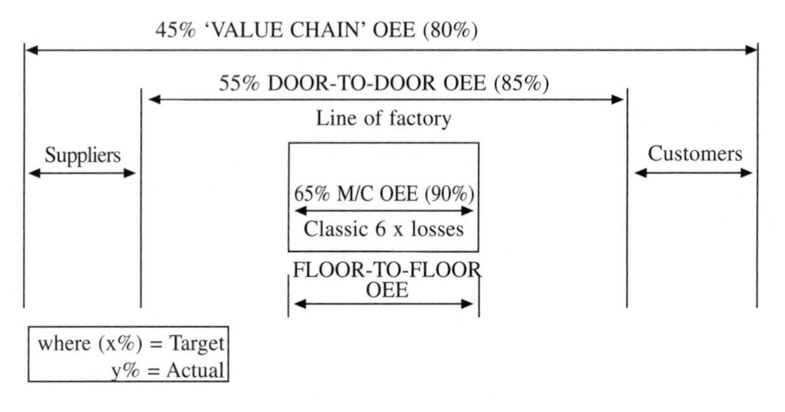

Figure 8.8 The value stream and the OEE

(ease of use), maintainability (ease of maintaining), reliability and safety. This includes labour and other operational resources which do not reduce when consumed (in the short term). These resources when released also have the potential to add value and improve competitive capability.

Transformation losses

Energy costs do not vary directly with output. In some businesses, 80 per cent of energy costs are fixed. As a result, such costs can be volume-driven. It is not just a case of switching lights off. Reducing minor stops through improved asset care will reduce energy losses while idling. Leaking air lines, once refurbished, will reduce electricity costs. Tooling care and design can also have a major impact on energy costs.

Maintenance materials also do not vary directly with volume. This is affected by factors such as levels of contamination, stop/start production, corrosion and brittleness as well as training, variation in production methods and, of course, human error. These costs reduce when not consumed.

Material losses

Often equivalent to 50 per cent of sales value, product design, improving process capability and improved working practices can all impact levels of material loss.

Management losses

The remaining value chain losses influencing this cover the company response to customer expectations (see Figure 8.9). For example, if current OEE results in a cost of £2.10 per unit as shown in Figure 8.10, the potential cost per unit at 10 per cent improved OEE is £2.00.

If the additional capacity cannot be sold, management will need to restructure overheads to compete with this achievable unit cost. Labour reduction, even

Figure 8.9 Customers drive our business

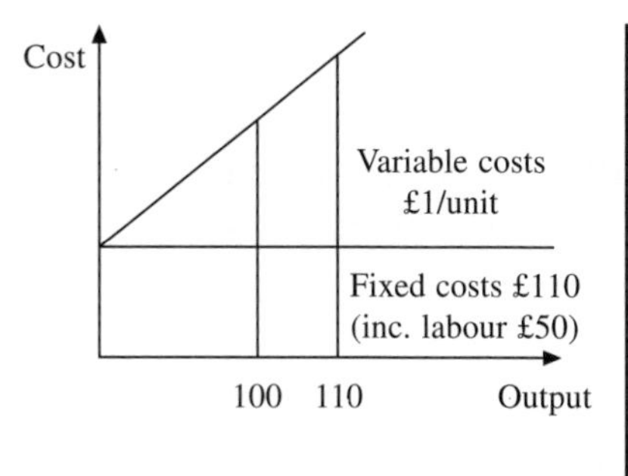

	OEE	80%	Produce more 88%	Produce same 88%	
(A)	Output	100	110	100	
(B)	Fixed Costs	110	110	105*	
(C)	Variable costs	100	110	100	
(D)	Total cost	210	220	205	
(E)	Unit cost D+A	£2.10	£2.00	£2.05	(D+A)
(F)	Contribution	1.05	1.15	1.10	
(G)	Unit sales price	3.15	3.15	3.15	
(H)	Total contribution	105	126.5	110	(F×A)
(I)	Return on capital employed	–	+20%	+5%	

In the example

- Producing 10 per cent more in the same time increases return on capital employed by 20 per cent
- Producing the same in 10 per cent less time increases return on capital employed by 5 per cent (*reduced labour cost by 10 per cent)
- What is your organization's potential return on capital employed?

Figure 8.10 OEE/loss relationship

if it is possible, will not be enough. This will reduce unit cost to only £2.05. There are other hidden losses associated with redundancy – not least the barriers it presents to continuous improvement. Finding the additional demand will avoid the loss of 5p/unit.

Loss prioritization

Loss modelling allows a comparison of potential cost structures at current and forecast OEE levels and volumes. Using best of best and average OEE improvement curves, it is possible to predict forward the likely cash flow gains from improved OEE. These areas of loss avoidance can be both linked to the appropriate TPM techniques and allow resources to be focused and then deployed through the pillar champions to the shopfloor teams.

8.4 Steps to achieve the TPM vision

Experience of implementing TPM has shown that the route to world-class performance begins with eliminating *sporadic* losses. Once these are under control, the task of eliminating *chronic* losses is made easier.

There are two main contributors to sporadic losses:

- equipment condition
- human error

Using TPM techniques, the route to addressing these factors takes around three years and is the main focus for the first two milestones of the master plan:

- Get everyone involved.
- Refine best practice and standardize.

This is the structured mechanism for the TPM master plan which integrates the vision and actions of the management team, providing consistent:

- prioritization
- expectations
- reward and recognition

As a result, skills are progressively developed so that:

- operators become technicians;
- maintainers become engineers;
- supervisors become managers;
- managers become entrepreneurs.

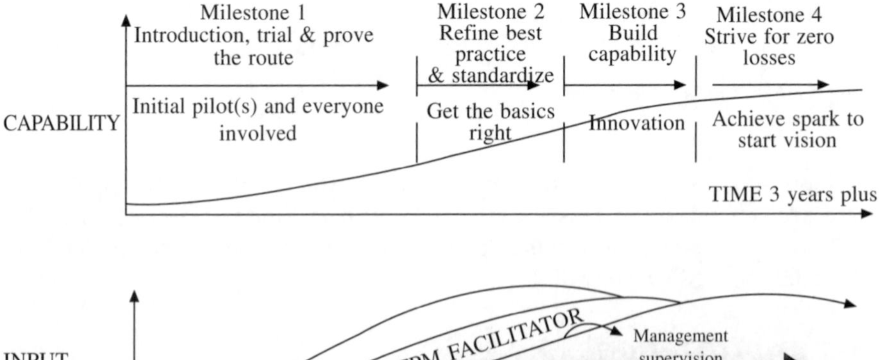

Figure 8.11 Skill development

The focus for facilitators, production, maintenance and management to deliver the TPM vision through the four milestones is set out in Table 8.3.

8.5 Management structure and the roles of supervisors

A plant-level TPM project structure taken from an actual case history is shown in Figure 8.2. This shows clearly the involvement of top management and the relationship of all the aspects of TPM which have been addressed in earlier chapters.

The roles and responsibilities of the management and supervisors as key contact to the TPM teams are repeated here to underline their importance:

- Input and release of people for TPM training
- Release of equipment for restoration and subsequent asset care
- Technical and historical information

Table 8.3 Steps to achieve the TPM vision

Introduction (pilots)	*Refine best practice and standardize (roll-out)*	*Build capability (promotion and practice)*	*Continuous improvement (stabilization)*
Facilitator-driven	Facilitator/supervisor-driven	Supervisor-driven	Small-group maintainer/operator-driven (i.e. self-sustaining
Production focus Selected operators in pilot area work on improvement plan	All departments begin improvement activities	All employees use TPM concepts Establish standardization	Autonomous small groups implement activities Early problem detection/solving
Maintenance focus Selected maintainers work with pilot team	Maintenance organizes to support plant-wide projects and planned maintenance	Training in maintenance skills for operators and maintainers	Reducing equipment lifetime costs
Management focus Policies and structure to support long-term commitment	Stimulating interest, managing resources	Encouraging teamwork, training and skill development	Striving for world class performance

- Plant and line OEEs
- Commercial and market benefits definition
- Visual management of information
- TPM publicity and awareness communication
- Inter-shift communications
- TPM activity logistics and facilitator support
- Standardization of best practice
- Spares forecast and consumption rates
- Hygiene and/or safety training and policy
- Input to problem solving and solutions

A dramatic message for managers and supervisors is embodied in Figure 8.12.

Requirement
To be a world-class manufacturer

Through
Just in time, lean production, one-piece flow

Prerequisite for success
We will be planning for failure regarding our daily and weekly output schedules unless the six losses of:

- Breakdown
- Set-up and adjust
- Idling and minor stoppages
- Reduced speed
- Defects and rework
- Start-up

are systematically tackled and eliminated through the TPM improvement plan. This will be reflected in an OEE performance of 85 per cent + rather than the c.70 per cent OEE of today.

The choice?
Either 100 per cent sustained commitment from management and supervision for TPM

or

Continue to plan for failure

Figure 8.12 The choice for management

8.6 Barriers to introduction of TPM

Inevitably, when major changes in an enterprise are being introduced there will be suspicion and opposition: this has been discussed in Chapter 7. Some of the common reasons for suspicion are as follows:

- Management shows impatience for quick fixes rather than 'stickability' and commitment.

- I operate, you fix: I add value, you cost money.
- Operators are taking our jobs away (say maintainers).
- TPM is a people reduction programme (threat of job losses).
- TPM is just another cost reduction driven programme.
- TPM is a hidden agenda to get operators to do the maintainer's job.

In the early stages, if communication is poor, then resistance will be inevitable. An open and clearly thought out agenda for TPM is absolutely vital.

8.7 Visibility of information

The importance of visible sources of information to reinforce discussion and verbal instructions cannot be too strongly stated. One of the major lessons learned in the early stages of TPM introduction in Japanese enterprises was the use of sight as part of communications to complement hearing. Some examples are as follows.

- *TPM equipment or activity boards* Handwritten documentation of status, progress and achievements prominently displayed in the work area, with objectives included. A schematic example is shown in Figure 8.13.
- *Black museums* Examples of problems solved, or waiting to be solved.
- *Training records* Displayed publicly; updated by trainees.
- *Notice boards* Located at factory entrances. Professionally implemented, with excellent graphics. Processes and achievements clearly analysed. Include safety records.

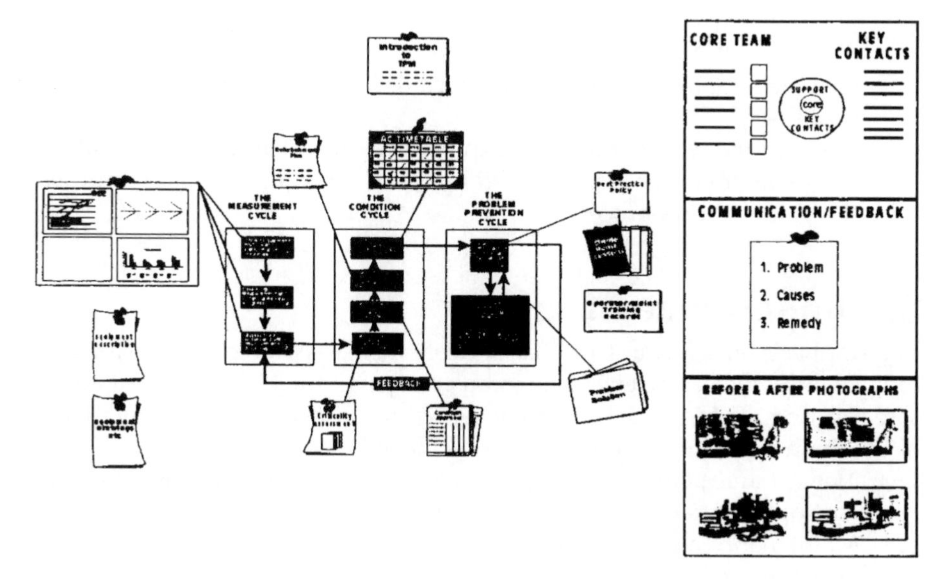

Figure 8.13 Lineside activity board

- *Operation information* A4 format located on or by the equipment.
- *Maintenance points* Marked by red arrows and frequency symbols. Colour coding for clarity of assembly/repair sequence. Optimum maintenance routes clearly marked. Machine defects tagged.

8.8 Support for teams and key contacts

Management support for these groups must be visible and total. Support of teams involves: active listening; supporting and, if differing, taking responsibility; stating issues which are your concern; and being specific. Some negative team practices which need to be gently corrected are: over-talking; not joining in; flying off at a tangent; and hijacking the discussion.

Some guidelines for helping teams to succeed are as follows:

- Agreed priorities and strategy
- Effective planning, control and delivery systems
- Clear organization of labour, equipment and materials
- Insistence on measurable results and individual commitment to them
- Encouragement to identify and meet task and process skills needs
- Active reinforcement of teamworking
- Promotion of a positive outlook to problem solving and new ideas
- Mutual trust
- Mutual support
- Good communication
- Shared objectives
- Managed conflict
- Effective use of skills

The role of key contacts is to:

- provide support to the team in specialized areas which impact on the shopfloor;
- improve the working relationship between direct and indirect staff;
- create an environment where all business functions can help improve the value adding process.

8.9 Importance of safety

The emphasis on safety at work has steadily increased in recent years, and in today's industrial scene everyone in an enterprise must be concerned about safe practices. TPM is very much concerned to enhance safe working. Some of the main ways in which this can be achieved are as follows:

- Neglect and penny-pinching are false economies in the context of the cost of injuries due to unreliable machines.
- Maintenance and safety are tied partners. Most injuries and accidents are caused by operators trying to intervene because their machines are not operating correctly.

- Maintenance means proper guarding, no exposed parts, minimum adjustment: it means the operator is protected.
- The Health and Safety Executive says most hearing damage is caused by badly maintained machines.
- When cleaning or driving our car we can identify at least 27 condition checks, of which 17 have significant road safety implications. Bring this good practice into work with you (see Figure 3.3).
- The notion of the competent and trained person, linked to assets that are fit for purpose and safe, plus statutory obligations, must be central to your TPM strategy, policy and practice.

8.10 Summary

Some of the intangible benefits of why TPM works are as follows:

- TPM is common sense and is therefore valued by employees and employers alike
- Practical vehicle for implementing the company's goals and vision
- Changes the employee's mind, creating ownership
- Belief in his/her equipment
- Protected and maintained by him/her
- Through self-help (autonomous maintenance)
- Give the employees confidence in themselves: create a feeling of 'where there's a will there's a way'
- Clean environment and environmentally clean
- Good corporate image

Figure 8.14 poses a question and provides the answer which epitomizes the TPM approach. Figure 8.15 illustrates what TPM meant to a team based on their experiences of running a sixteen-week TPM pilot exercise at an automotive manufacturer in the north of England.

There is no better way of rounding off this chapter than by quoting the general manager of the plant after attending a team presentation of a TPM improvement plan pilot:

> We started our TPM programme – or TPM journey as I prefer to think of it – about three months ago, so it's early days yet. However, the things that struck me most about the TPM team's

---

Question
If you haven't got the time to do things right the first time . . .
How are you going to find the time to put them right?

Answer
TPM gives you the time to do things right the first time, every time!

Figure 8.14 TPM: the answer to a problem

presentation today were their obvious enthusiasm for what is proving to be a grass roots process with real business benefits. The other factor which is quite clear to me is that TPM can only be sustained provided our supervisors and managers support the TPM process wholeheartedly. Our workforce obviously values the process: it is up to us to give them the time and full resources to carry it out. We've always known that our equipment and process capability is not what it should and could be. Everyone thinks about quality output. TPM adds the missing link: quality output from world-class and effective equipment.

Today	People	Matter
Totally	Pampered	Machines
Totally	Perfect	Manufacturing
Training	People	Meaningfully
Teamwork	Production	Maintenance
The alternative:		
Tomorrow?	Probably . . .	Maybe . . .

Figure 8.15 What TPM means to us

9
TPM for equipment designers and suppliers

Behind the plant and equipment used in the production process there are three functional groups, namely:

- Operations
- Commercial
- Engineering

These make up three essential partners for new product/equipment introduction. This chapter describes in outline how these activities must be co-ordinated and focused on the TPM objectives. The partnership requires a sustained drive towards improving project and design management performance through the elimination of hidden losses such as poor maintainability, operability, and reliability early in the equipment management process.

Designers and engineers need to improve their skills by:

- regular visits to the shopfloor and learning from what operators and maintainers have to say;
- studying what has been achieved in equipment improvement as a result of self-directed and quality maintenance activities;
- gaining hands-on experience with equipment, including operation, cleaning, lubrication and inspection;
- supporting P–M analysis as part of the key contact/team activities;
- conducting maintenance prevention analyses.

Figures 9.1 and 9.2 show how the five goals of TPM can be achieved through design feedback, early warning systems and objective testing of new ideas.

Figure 9.3 portrays the benefits of using TOM design techniques, TPM (D), as the driver for Early Equipment Management (EEM). All partners are involved in achieving the continuous improvement habit, learning how to deliver flawless operation in less time.

It also shows what the TPM (D) process can deliver over the life of the equipment. The gap between typical (or traditional) output/value and true potential by getting it right in the early stages is huge.

Figure 9.4 illustrates the concept that two-thirds of the lifetime costs of new equipment is determined (but not spent) in the early design specification stages and can, therefore, be said to be designed in. This serves to emphasize

Goal	Process			
Equipment effectiveness	Formalize	Standardize	Transform	Optimize
Autonomous maintenance	Define	Raise awareness	Practise with support	Autonomous activity
Skill development	Standardize	Improve	Transfer skills	Systemize
A planned maintenance system	Restore	Simplify	Stabilize	Extend
Early equipment management	Define	Design	Refine	Improve

Figure 9.1 Early equipment management: linking the five goals of TPM

the importance of getting the design right first time, not just for intrinsic reliability but also for fitness for purpose, operability and maintainability – and also, of course, safety and environmental issues.

As outlined in the following sections, there are three major TPM (D) techniques which promote close collaboration between the three essential partners:

- Objective testing
- Milestone management
- Knowledge base management

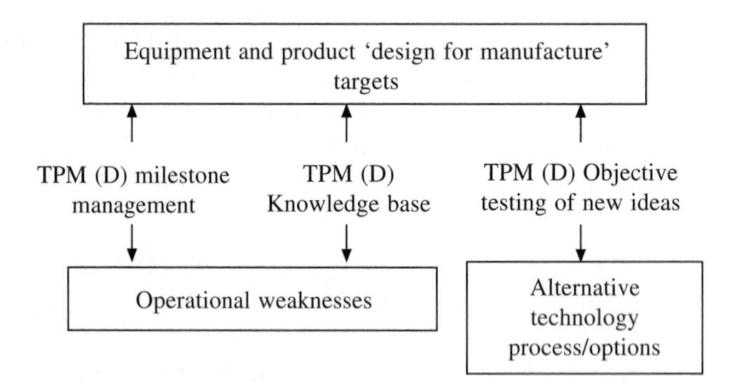

Figure 9.2 Early equipment management: framework for maintenance prevention

9.1 Objective testing

This is technology/process design-oriented and requires a search for new ideas using:

- *Intrinsic reliability* Repeatability of optimum conditions; simple construction; simple installation

- *Operational reliability* Tolerance to conditions; simple manipulation; ease of maintenance
- *Lifetime costs*

All this is a part of continuous improvement.

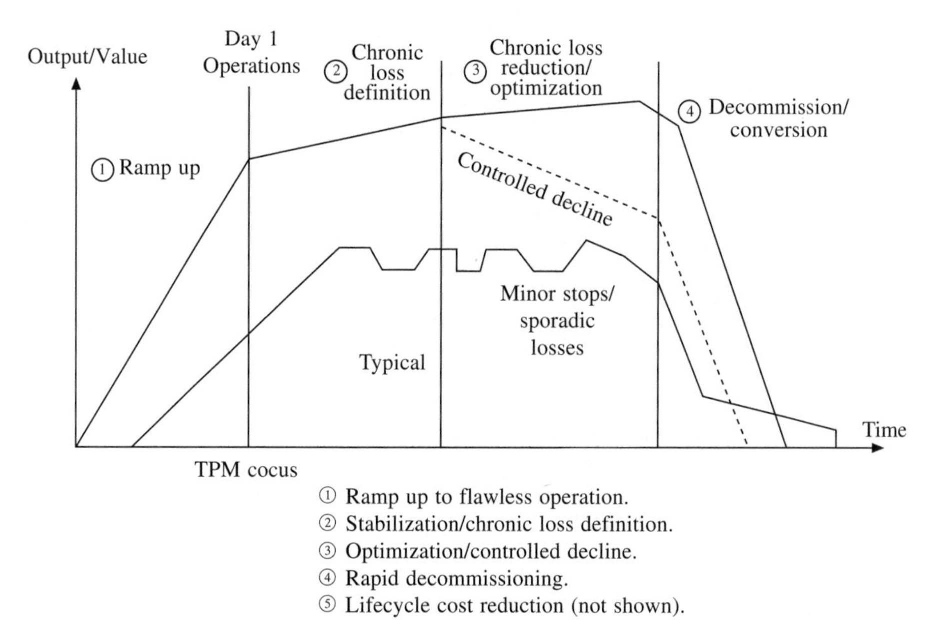

Figure 9.3 What can TPM(D) deliver?

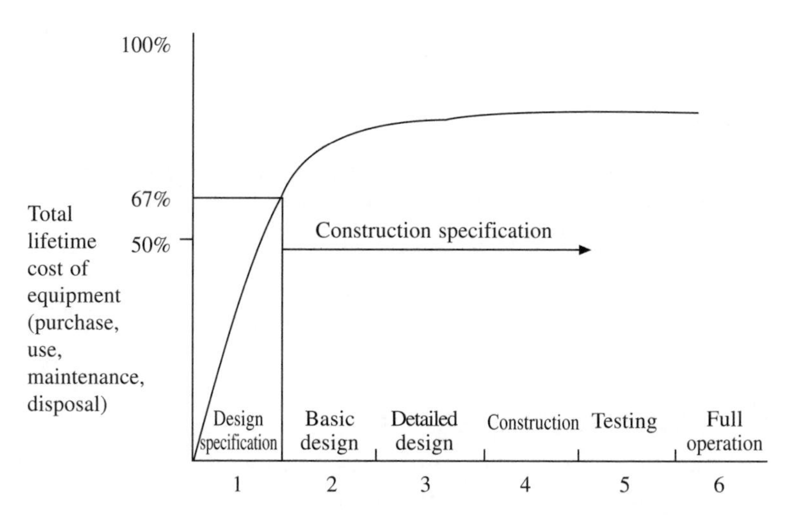

Figure 9.4 Early equipment management leads to greater cost control and flawless operation

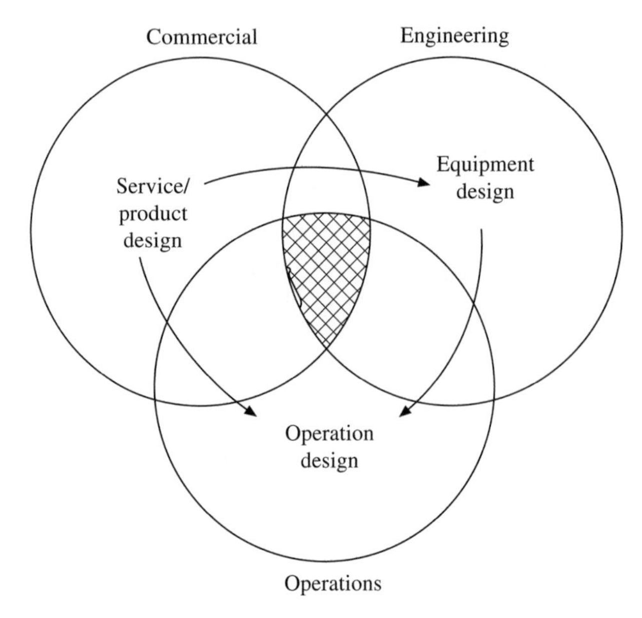

Figure 9.5 Design issues: product design influences equipment design, and operations design is influenced by both

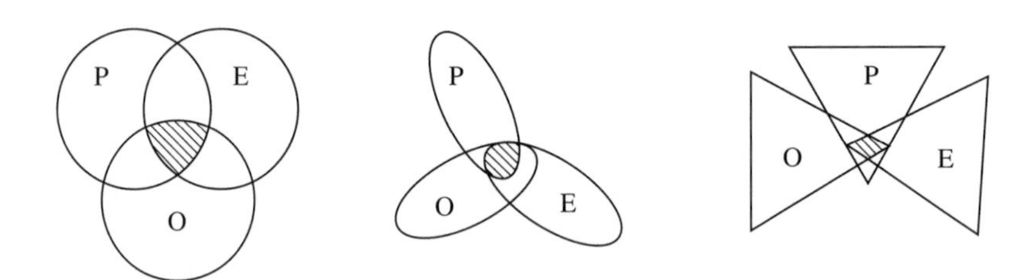

Figure 9.6 Selecting the best design: many possible combinations of product, equipment and operation design. Customer requirements for timely, high-quality, low-cost products and services must provide the basis for selecting the preferred option

9.2 Milestone management

The commercial, operations and engineering subteams each have a role throughout the design process, as set out in Figure 9.7. Milestone reviews aid early problem detection and secure buy-in at each stage.

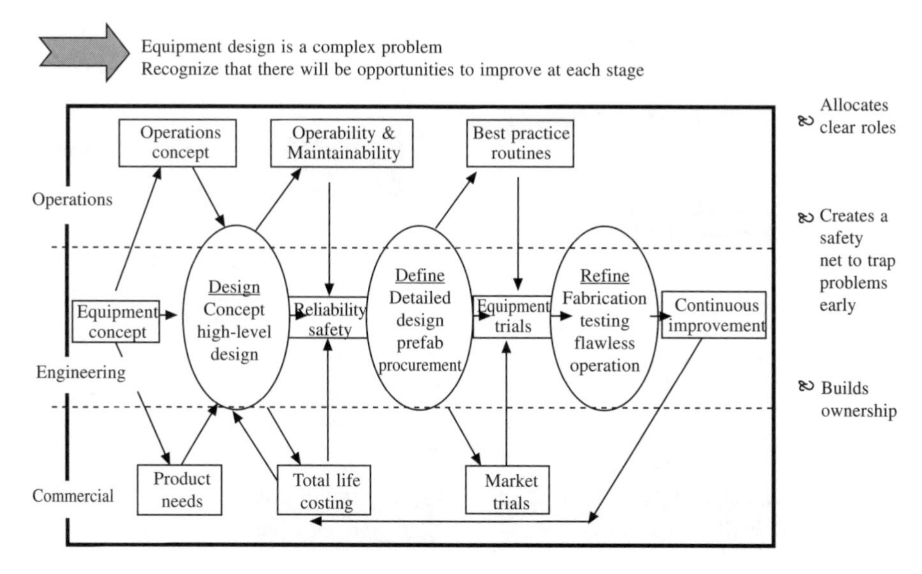

Figure 9.7 TPM(D) milestone management roles

9.3 The knowledge base

This requires study and analysis of:

- TPM activities and solutions: best practice routines, single-point lessons;
- operability: make it easy to do right and difficult to do wrong;
- maintainability: breakdown/inspection reports, maintenance prevention;
- reliability: defect analysis, six losses, OEE.

Figure 9.8 illustrates the links between objective testing and feedback in the knowledge base. Figure 9.9 illustrates a key knowledge base function to define reasons for defects and ultimately design out the weaknesses.

9.4 Refining the knowledge base

The achievement of effective knowledge base usage entails setting goals and determining measures which will progressively eliminate or simplify component parts.

Analysis steps to design out those defects include:

1 Collect breakdown analysis data and single point lessons issues
2 Analyse and ask 'why' five times:

- Consequences of failure?
- Causes of failure – human error?
- Improve reliability?
- Improve maintainability?
- Set and maintain optimal conditions?

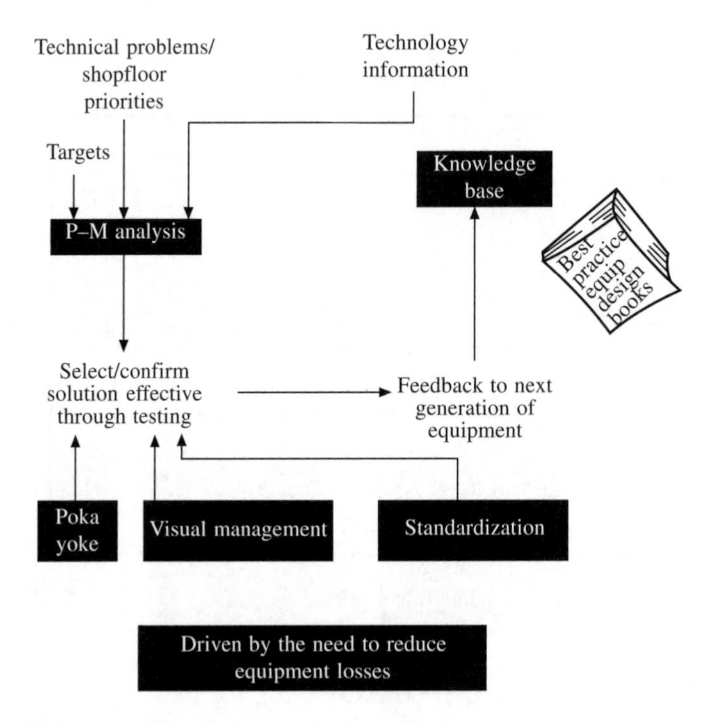

Figure 9.8 Objective testing

9.5 Standardization

Standardization is one of the main outputs from refining the knowledge base helping to deliver easy maintenance and trouble-free operation. Standardization can be applied to:

- operation procedures
- set-ups and changeovers
- asset care routines
- fixtures and fittings:
  - adaptors
  - connectors
  - thread sizes
  - screw, nut, bolt heads
  - quick release
- monitoring and control:
  - gauges
  - oil
  - heat
  - electric
  - pneumatic
  - instrumentation

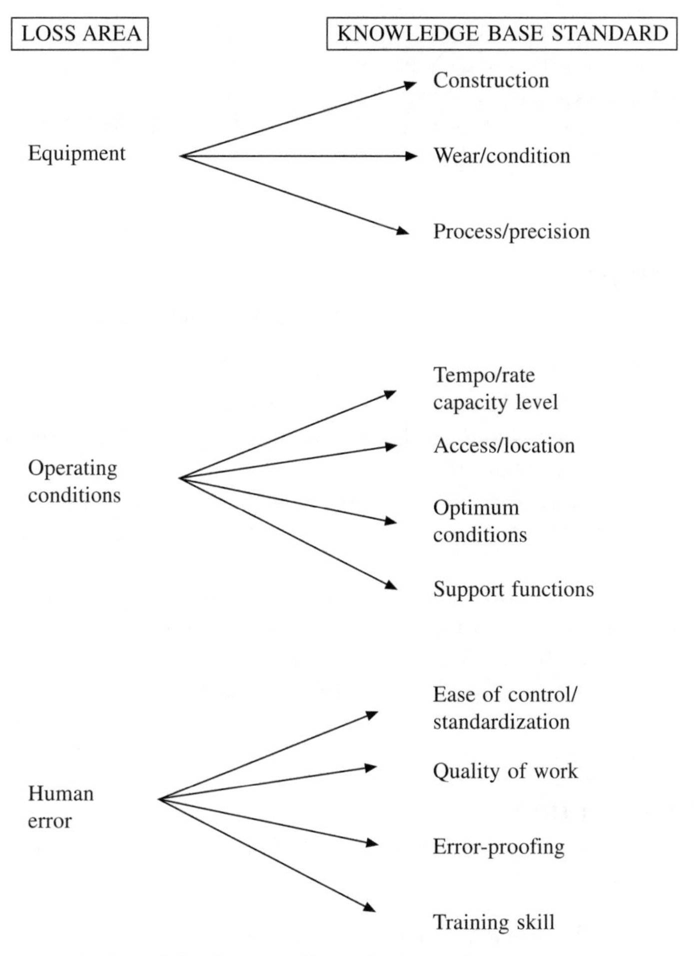

Figure 9.9 Inputs to the knowledge base to aid standards setting

9.6 Checklists

The knowledge base should contain checklists such as those below to guide progress through the TPM (D) milestones.

Effective design/specification

- Can the item of equipment be eliminated? (Is it vital to the process or as a result of the design?)
- Can the item or part be integrated with the adjacent part?
- Can the item be simplified? (Can it be a standard part rather than a special one?)
- Can we standardize the item with another item?

- Can the equipment/item cope with the environment? (dust/heat/damp/vibration: adverse as well as normal conditions)
- Can the equipment control be simplified?
- Can the item be made of a cheaper/different material?
- Can a cheaper service be used?

Operability

This is aimed at making it easy to do right, difficult to do wrong.

- Are frequent adjustments required?
- Are handles or knobs difficult to operate?
- Are any specialized skills or tools required for operation/adjustment? (start-up, shutdown)
- Are blockages/stoppages likely? (How are they resolved?)
- Has any diagnostic function been built in? (glass panels, gauges, indicators)
- Start-ups and shutdowns: is additional manning required?
- How robust is the equipment? (Will the equipment break down or product quality be affected by poor operation?)
- Is the operator's working posture unhealthy?

Maintainability

The keys here are to try to eliminate maintenance or to make it easy, infrequent and low-cost.

- Can we eliminate the need for maintenance?
- Are areas easy to clean, lubricate or check?
- How long is the equipment set-up time?
- How frequently does the equipment need tuning or calibrating?
- Are specialized maintenance skills required?
- Can failure be predicted?
- Have any self-diagnostic functions been built in? (Is it easy to find the cause of failures?)
- Can parts be easily replaced and plant restored quickly?
- How reliable is the equipment?
- Can we extend the maintenance interval?
- Does the equipment structure facilitate maintenance? (lifting heavy parts, etc.)
- What routines are required?
- What spares support is required?
- Can breakdowns be restored cheaply? (Can spare materials and parts be purchased cheaply?)

9.7 Typical equipment design project framework

The core project team should include representatives from the three essential partners as shown in Figure 9.10. This could be made up of:

- designer/specifier
- planner/specifier
- manufacturing engineer
- equipment operator
- equipment maintainer
- equipment supplier
- facilitator

The key contacts could include:

- purchasing
- finance
- quality
- product engineering
- process engineering

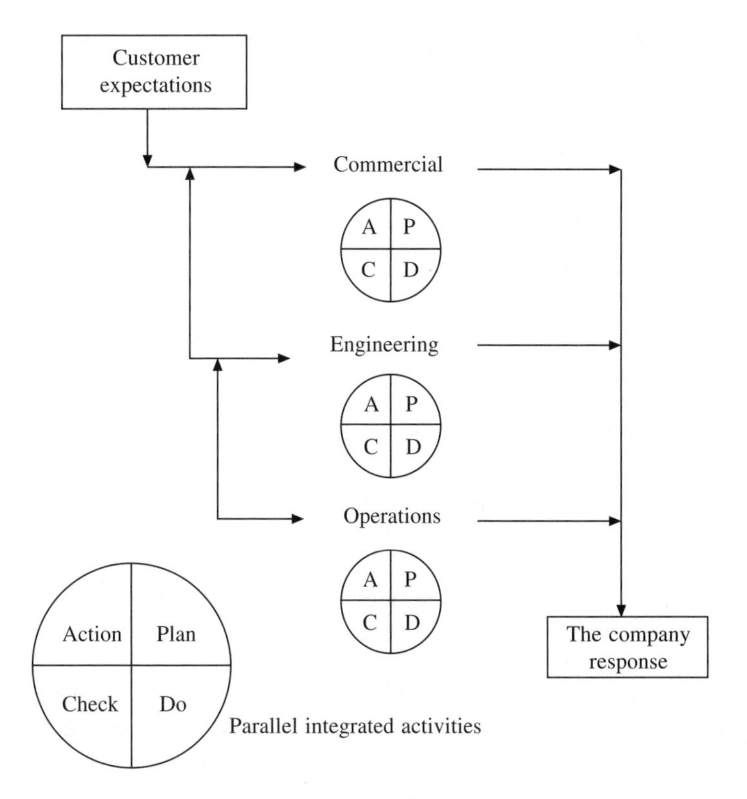

Figure 9.10 Delivering world-class performance

A timetable for an equipment design project is shown in Figure 9.11. Initial training would involve the core team and the key contacts. The activity sessions are described below.

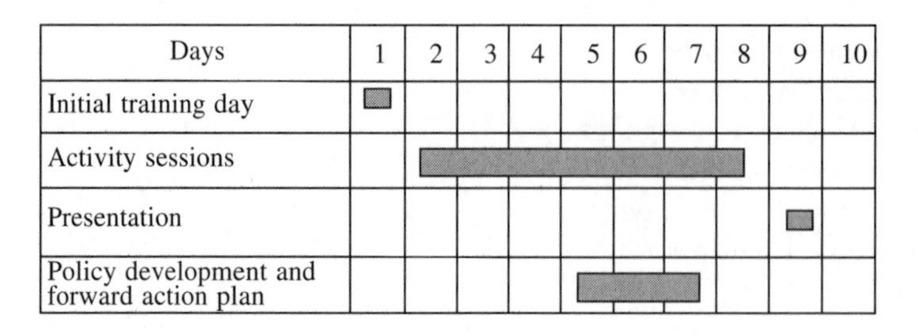

Figure 9.11 Equipment design project timetable

Activity session exercises

Setting design targets

- critical ratings
- design efficiency
- operational conditions
- process trade-offs
- intrinsic reliability
- operational reliability

It is highly desirable to involve the equipment supplier at this stage.

Define

- Confirm trade-off analysis of basic outline on a modular basis.
- Establish testing/audit criteria for each module/subsystem.
- Conduct outline critical assessment to predict equipment weaknesses per module.
- Establish criteria for standardization of components/spare parts.

Design

- Confirm/refine each module/subsystem at a detail level.
- Incorporate error-proof devices for flawless operation.
- Establish tooling/maintainability criteria.
- Establish asset care regimes with supporting visual management.
- Simulate cleaning and inspection activities to improve operability.
- Simulate maintenance activities to improve maintainability.
- Feed back improvement suggestions.

Refine

- Review construction constraints/opportunities.
- Agree quality audit milestones for main construction process.
- Define detailed project plan.
- Establish how the equipment will be located in relation to other equipment (layout considerations).
- Complete quality audit reviews.
- Establish best practice routines and develop training material.

Trial/testing

- Project planning.
- Installation, including workplace organization and OEE measurement
- First-run trials.
- Confirm best practice and standardization.
- Joint sign-off of operation.

Improve

- Maintain normal conditions.
- Stabilize best practice routines.
- Strive to establish optimum conditions.
- Deliver better than new performance.

Conclusions

Figure 9.12 shows how the nine-step TPM improvement plan may be used to provide inputs to the knowledge base. Figure 9.13 shows how the improvement plan can be used to aid TPM (D) milestone management.

Companies who adopt the philosophy of TPM for design will have the potential for a huge commercial advantage resulting from equipment with minimum total life cycle costs, which delivers high overall equipment effectiveness levels and flawless operation.

• Equipment history	Record of reliability
• OEE	Trend indicates need for action
• Six losses	Record of areas of improvement
• Criticality assessment	Formal review of design performance post-installation
• Condition appraisal	Audit/record of deterioration
• Refurbishment plan	Record of life time costs
• Asset care	Planned maintenance costs
• Best practice routine	Activities need to achieve flawless operation
• Problem solving	Opportunities to pass on lessons learned

Figure 9.12 Using the improvement plan as inputs to the knowledge base

	Design concept	Basic design	Detailed design	Build/ install	Testing/ refine	Implement/ use
Equipment history			✓	✓	✓	✓
OEE	Set targets	Assess trade-offs	✓	✓	✓	✓
Six losses	Setting zero targets				✓	✓
Critical assessment		✓	✓			✓ Support training
Condition appraisal	Feedback on weaknesses				Set standards	✓
Refurbishment plan	Assessment of lifetime costs					
Asset care		✓	✓		✓	✓
Problem solving	Target setting			✓	✓	✓
Best practice routines					Aim for flawless operation	✓

Figure 9.13 Using the TPM improvement plan in the design process

10
TPM in administration

10.1 An overview

Developed from its original well-proven roots of Total Productive Maintenance, Totally Productive Operations (TPO) looks at the complete value stream. Key components of this are TPM (Total Productive Manufacturing) and TPA (Total Productive Administration), as shown in Figure 10.1.

The value stream is driven by customer demand, and hence we have to maximize value across the chain.

Our goal is to maximize added value by eliminating waste 'in all that we do'. In the case of TPA we might add the phrase 'in support of our customers and core business processes'.

Some typical, but not exhaustive, application areas are shown below:

- Management Information Systems
- Finance
- Purchasing
- Human Resources
- Stores/Warehousing
- Quality Assurance & Control
- Sales Force Activity
- Order Processing
- Design & Engineering
- Research & Development
- Project Management
- Marketing
- Training & Development
- Production Scheduling
- Despatch and Delivery

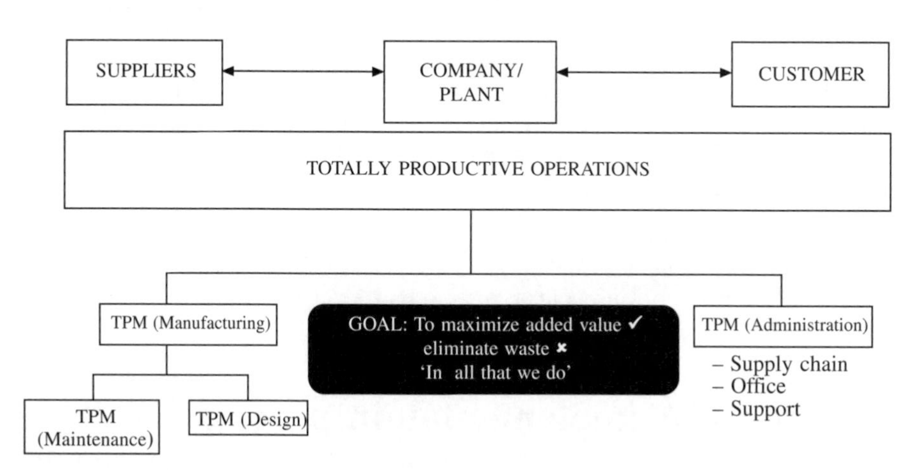

Figure 10.1 The value stream and TPM

The application of TPM in Administration, or TPA, has parallels with the approach used in manufacturing. Many administration problems are unmeasured and therefore hidden, just as they are in manufacturing. This chapter looks at the application of TPA. The issues are just as relevant to non-manufacturing industries such as construction, where the workplace is not fixed and logistics/planning has to deal with this added dimension. It can also be applied to computer-based and financial services work environments, where CAN DO is as important as ever.

The wide variety of tasks carried out by administration makes it appear complex and difficult to standardize. Therefore, when there are peaks in workload it can be difficult to know how to smooth out the bottleneck.

As a result, there are often minimal standard practices, formal training, few, if any, single-point lessons, fool-proofing or systematic loss elimination activities. Individually, good administrators are excellent organizers, but this is usually limited to their immediate work rather than the system as a whole. Typically, administration systems are characterized by the weaknesses shown in Table 10.1.

Table 10.1 Administration issues and weaknesses

Issue	*Weakness*
Dependent on individual initiative	Difficult for others to fill in
Much manual and discretionary work	Difficult to learn from experience
Numerous records and ledgers to be maintained	Duplication of documents, files and information
Current job processing status is difficult to assess	Difficult to measure progress or to improve quality standards, productivity or delivery performance

TPA uses the CAN DO workplace organization steps to address the office infrastructure, i.e. filing systems and layout issues.

In parallel, office systems are reviewed using the improvement plan phases as shown in Figure 10.2 of:

- Measurement cycle
- Development cycle
- Problem prevention cycle

10.2 The TPA implementation process

The TPA implementation process is illustrated in Figure 10.3. It comprises the 'planning' or scoping stage, followed by the implementation phase.

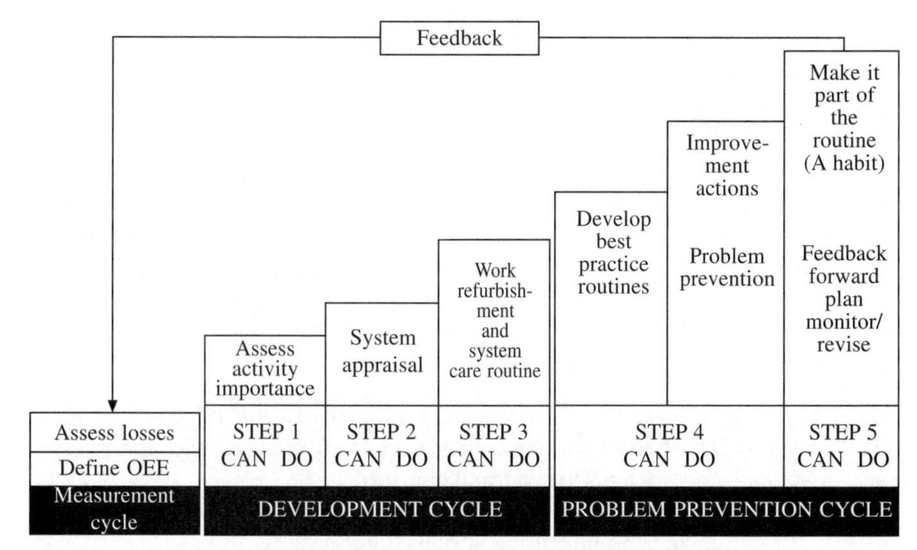

Figure 10.2 TPA improvement plan

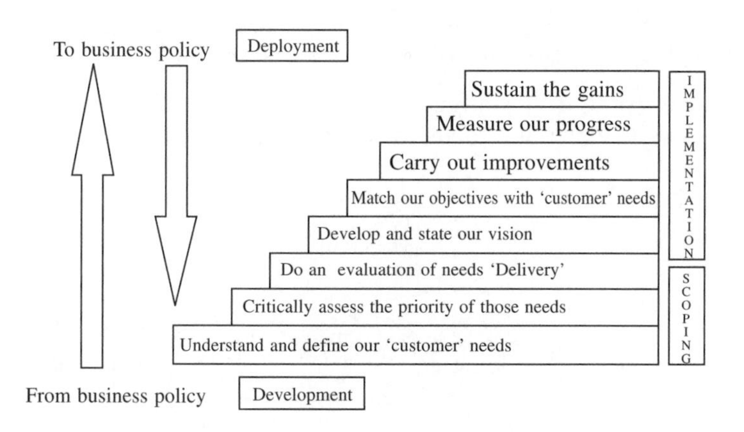

Figure 10.3 Overall TPA process

The essential TPA policy leadership includes:

- Understanding 'customer' needs
- Evaluating current loss areas
- Matching customer/internal improvement priorities
- Improvements techniques to bridge the gaps
- Measurement of progress against hidden loss targets
- Holding the gains in value from improved administration

Each direct and indirect support function needs to 'fit' with the overall

TPA policy development to support deployment of that policy and maximize value adding activities.

Scoping study

The scoping study is used to raise awareness of TPA, to agree priorities for action, issues to overcome (both hard and soft) and a timetable for the pilot. The priorities for action include an assessment of current and future customer needs (internal and external) and how well the function is meeting those needs. Typically, this involves establishing data collection to measure current levels of Overall Administration Effectiveness (OAE) losses.
The areas to concentrate on are:

- Vision development: How does the administration function support the delivery of external customer needs and business benefits? What losses can be addressed by this function?
- Infrastructure: Who is the TPA pillar champion, who are the key contacts and facilitator, who should be given a general awareness?
- Team profile: Should it involve customer/supplier departments?
- Pilot timetable mapping out bottom-up and top-down activities
- Roll out concept to systematically involve other functions
- Development of a TPA 'Spark to Start' vision

Figures 10.4, 10.5, 10.6 and 10.7 show typical examples of:

- a business unit TPA infrastructure
- typical 'key activity' focus for the OAE champion

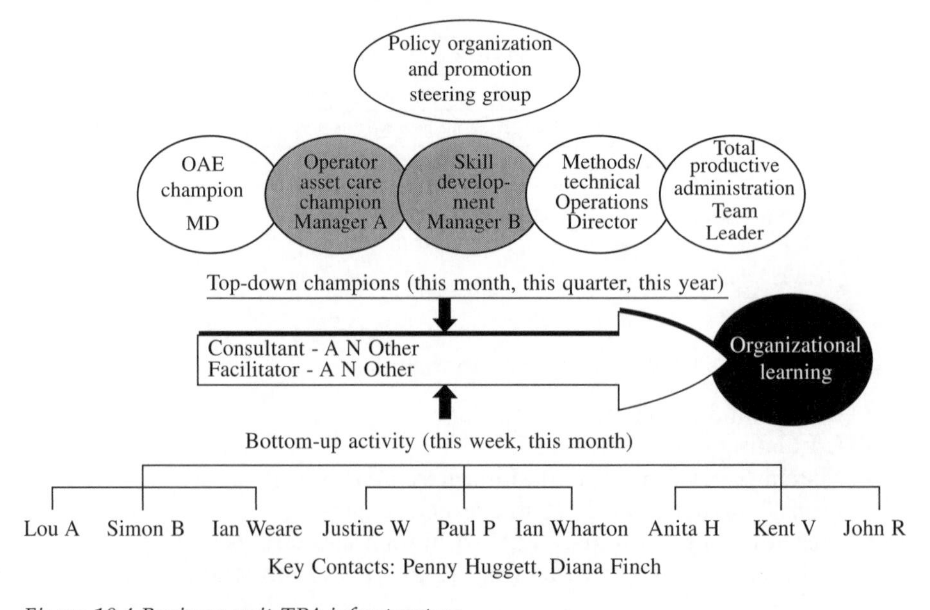

Figure 10.4 Business unit TPA infrastructure

KEY ACTIVITY: *To ensure that overall administration effectiveness has quantifiable current, BOB and CI targets*

SHORT TERM (DAILY/WEEKLY)	To check with TPA champion that blockages/breakdowns/overloads are being overcome by telephoning daily and 1 hour review meeting each week (PA9) 
MEDIUM TERM (MONTHLY/QUARTERLY)	Align TPA activity team's priorities with business needs – especially '100% customer ready service'/elimination of waste and TPA champions PA's 2, 8b, 8c, 9 and support of PA7 and 8a 
LONG TERM (ANNUALLY)	Review plans bottom-up and integrate with top-down business goals 

Figure 10.5 OAE Pillar Champion key activities

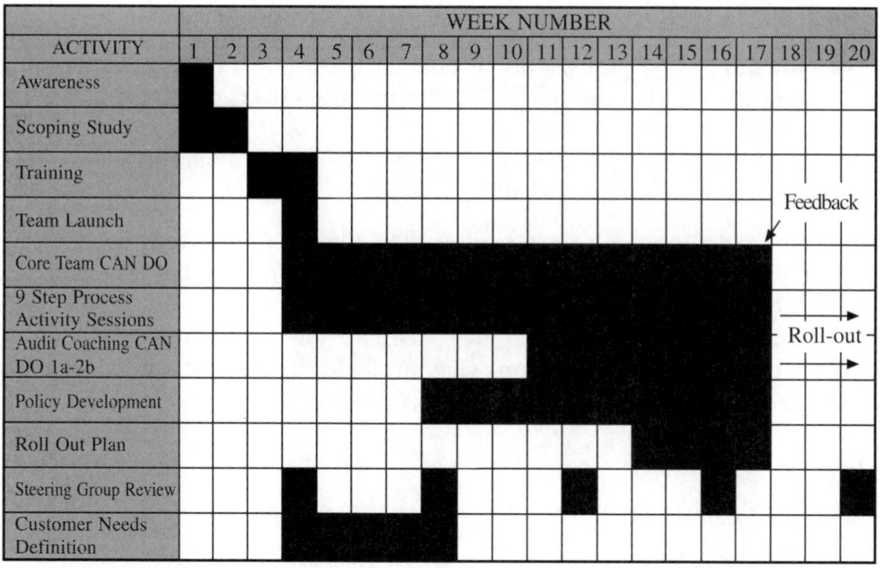

ACTIVITY	WEEK NUMBER 1	2	3	4	5	6	7	8	9	10	11	12	13	14	15	16	17	18	19	20
Awareness	■																			
Scoping Study	■	■																		
Training			■	■																
Team Launch				■																
Core Team CAN DO				■	■	■	■	■	■	■	■	■	■	■	■	■	■			
9 Step Process Activity Sessions				■	■	■	■	■	■	■	■	■	■	■	■	■	■			
Audit Coaching CAN DO 1a-2b											■	■	■	■	■	■	■			
Policy Development								■	■	■	■	■	■	■	■	■	■			
Roll Out Plan														■	■	■	■			
Steering Group Review				■				■				■				■				■
Customer Needs Definition				■	■	■	■	■												

Feedback

Roll-out

Figure 10.6 TPA typical timing plan

- a typical TPA activity timing plan
- a TPA 'Spark to Start' vision

Figure 10.4 shows the essential TPA infrastructure of:

- the Steering Group, comprising pillar champions, facilitator and consultant, who meet monthly

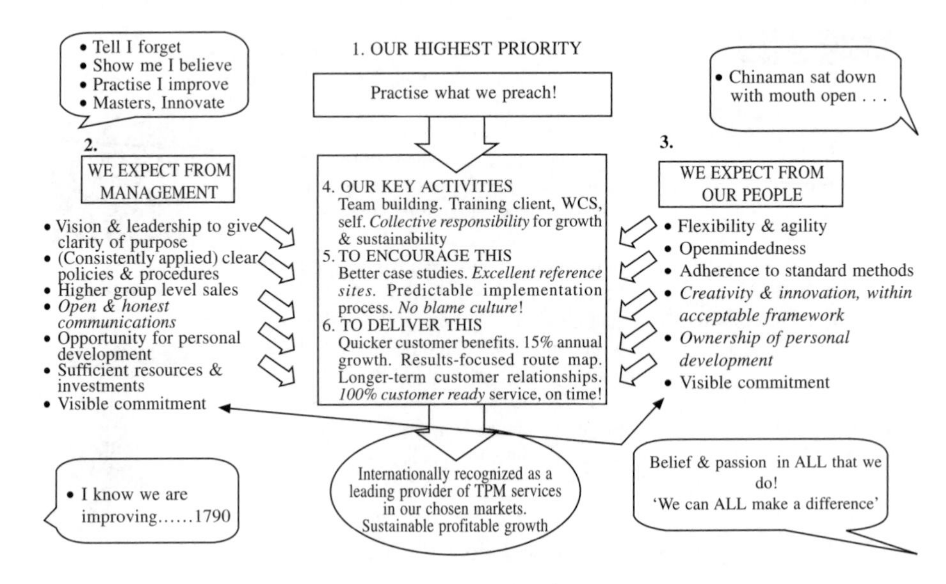

Figure 10.7 TPA 'Spark to Start'

- the pillar champions, who provide the direction in terms of developing and then deploying the policy (for example OAE, as shown in Figure 10.5)
- the bottom-up TPA team improvement activities, which are concerned with this week/this month

Figure 10.6 shows a typical timing plan for the first twenty weeks of a launch programme, including the activity sessions for the initial pilot. Further details are included in Table 10.4.

Activity sessions usually take place once a fortnight, maximum eight hours, minimum four hours per session. They are held 'on the job' and comprise a multi discipline core team, plus their facilitator and key contacts as invited.

Pilot

Following the scoping study, a pilot is used to:

- gain experience of using TPA techniques and principles;
- identify road blocks to progress;
- develop a model to convince others;
- confirm the potential of TPA to reduce wasted effort and improve customer service;
- establish a realistic and achievable roll-out plan for the rest of the administration functions.

Milestone 1 Introduction (Everyone involved)

The roll-out cascade provides a systematic route to involve all administration functions in the use of TPA. The sequence of roll-out should reflect the business priorities. Here the main emphasis is on reviewing and formalizing existing systems and processes. Work scheduling is also formalized to make it easy to track work progress against planned service level criteria. The aim is to minimize the tasks required to maintain the administration systems and to ensure that areas of hidden loss are at least identified and prioritized so that the available time can be directed at the most important.

A key goal at this stage is to reduce retrieval time for all information to 30 seconds or less. Activities at this stage will be primarily directed at internal processes and allocation of responsibility for shared files and common areas.

Visual focus: labelling and organization will make it easy to do things right.

Milestone 2 Refine best practice and standardize

The milestone focus is rationalizing and minimizing administration tasks. There is a shift away from individual to group tasks so that as the workload fluctuates, the load can be shared.

A key goal here is to reduce filing space and, therefore, put-away time as well as lead times and processing stages. Activities may involve close co-operation with internal/external customers and suppliers.

Visual focus: work co-ordination techniques to highlight potential work overload and the need for reallocation of resources.

See Table 10.5 for more details.

10.3 Applying TPA

Below is a suggested programme for TPA pilots. This is intended to highlight differences and assist the experienced TPM practitioner to operate in the administration environment.

The TPA approach uses CAN DO and the TPM improvement plan techniques in a similar way to TPM in manufacturing. There are, naturally, changes of emphasis. Some principles are directly applied, such as:

- restore before improve as a route to current system restoration and understanding of the administration systems;
- the definition of routine activities and roles based on the need for technical judgement;
- the use of a pilot to learn the lessons prior to roll-out.

Other activities are uniquely administration oriented, such as:

- the move from individual-based to group-based activities. This permits the reallocation of resources to overcome workload problems;
- the use of visual methods to organize and progress work as well as highlight backlogs;
- reduction in filing space by 50 per cent;
- retrieval of routine information within 30 seconds or less.

The emphasis on cleaning equipment becomes an emphasis on ease of checking/archiving data (both physically *and* electronically) and reducing effort required to maintain the workplace. Here a major source of contamination is excess or out of date paperwork. The equivalent to reducing sources of contamination is the use of a 'One is best' campaign, e.g. one-page memos, one copy filed, one-hour meeting, and so on.

As set out above, during Milestones 1 and 2, this is applied within the department. During Milestones 2, 3 and 4, this is spread out to include internal/external customers and suppliers.

As with TPM, the first two milestones focus on a move from reactive to proactive management by standardizing core competences. This releases experienced resources to develop added value services to aid the competitive position of the business.

The reason this is referred to as TPM in Administration rather than TPM in the Office is that the process has been found to be effective in non-office environments such as stores, warehousing and support functions which do not have a fixed location (see 10.1 – An Overview).

Team launch

The initial training activity is similar to a conventional core team launch. The important point to emphasize here is the fact that hidden losses occur in any environment. TPA is a tool to trap and progressively eliminate these dynamic system losses.

Core team members need to feel comfortable with the distinction between the following loss categories (Table 10.2).

It is useful to explore with the team how the six losses can be expressed in relevant administrative terms. Although there can be more categories of loss, these six provide a framework for systematically addressing the main sources of waste. The team should agree how to capture the level of losses in each case to develop and monitor overall administration effectiveness.

Table 10.2 Loss categories and examples

Loss category	*Definition*	*Loss examples*
Availability (improve information collation/retrieval)	Things which prevent the job from starting	Inaccurate records (system breakdown)
Performance (improve job process stages)	Things which extend the processing time	Cross-referencing not correct (minor stop) Lost file (reduced speed loss)
Quality (reduce risk of human error)	Things which influence the quality of the work	Misdirected document (rework) Initial errors (start-up loss)

An example of using the 'iceberg' approach and then categorizing the losses as part of the measurement cycle is shown in Figures 10.8 and 10.9 respectively.

During the launch the team should define their vision of the future: 'If you walk into the area in three years' time, and we have been successful at

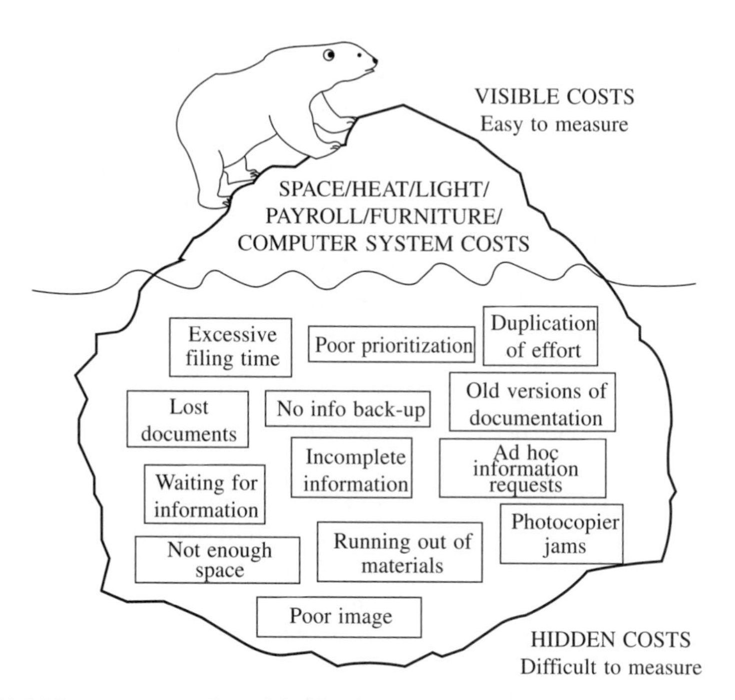

Figure 10.8 Measurement cycle tool: hidden losses

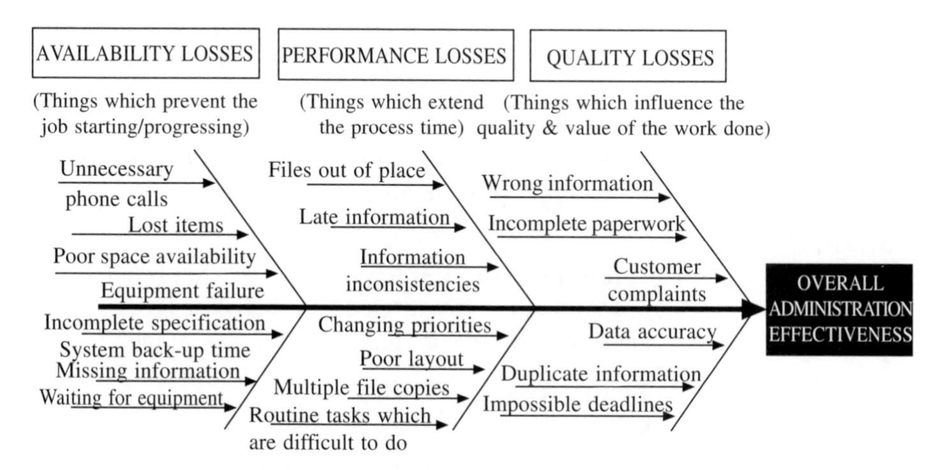

Figure 10.9 Measurement cycle tool: typical loss categorization

eliminating most of these losses, what would we expect to see in physical terms?' How will we know we have improved? How will working practices have been changed? (See also Figure 10.7.)

Consider those factors which help and hinder progress towards the vision. List those people who need to be involved to overcome the 'hinders'. (These will become key contacts.)

Discuss with the team meeting times and dates. Ideally, meet two weeks out of three for full-day team sessions, if possible. In small offices, only half a day may be possible. This may need to be fitted in around month-end procedures, but it should provide quality time for the team to release themselves from the day-to-day pressures.

The team should also plan routine CAN DO improvement activity. This may require allocation of areas to two or more individuals. It may also involve some prioritization. Try to gain agreement to a routine daily 15 minutes' activity. A weekly half-day would also be an alternative, but eventually we would expect to be able to progress towards a daily improvement clear and clean regime, with the emphasis on reducing the time required to maintain workplace organization.

Finally, the team should decide where to locate their TPM board and take 'before' pictures as evidence of improvement and to sustain future motivation.

Activity sessions *(see Table 10.4)*

Prior to core team activity sessions, record current losses and allocate CAN DO areas. Begin the big clear and clean. Activity sessions follow the format set out below. Sessions should be based on around 50 per cent briefing/ analysis and 50 per cent practical activity. If only half-day sessions are possible, the practical activity can be planned for a separate day. The following provides information to support the TPA improvement plan process steps.

Session 1

Carry out an outline brown paper modelling exercise to assess filing, numbering and labelling systems. Extend the clear and clean to filing systems and introduce routine clear and clean activities to break the back of the clear and clean task (e.g. 15 minutes per day). See Figures 10.10 and 10.11.

Focus on preventing unnecessary items from entering the areas to ensure that everybody understands the change is for good.

Session 2

Introduce the CAN DO audit to confirm progress/status. Don't move on to CAN DO before the required level of discipline has been achieved.

Figure 10.11 shows the first and second steps of CAN DO, namely:

Step 1: getting rid of everything unnecessary
Step 2: creating a right place for the things you need

TPA is about workplace organization using the CAN DO philosophy: making it easy to do things right ✓ and difficult to do things wrong ×.

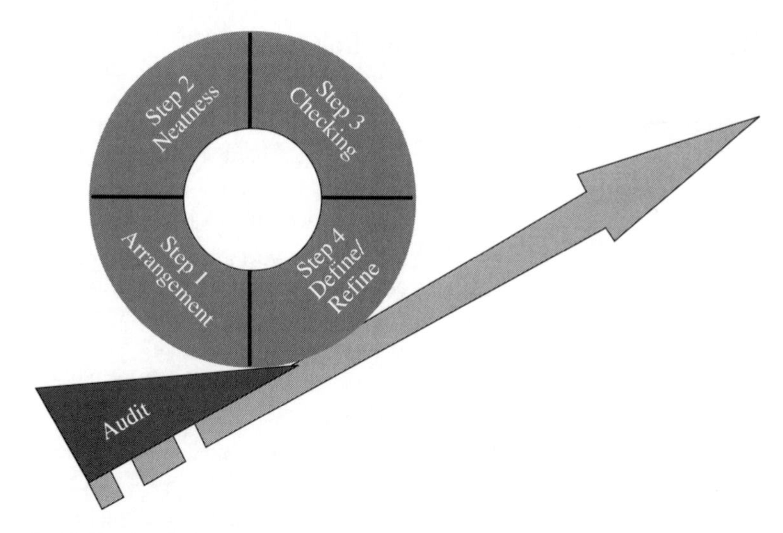

Figure 10.10 CAN DO in TPA

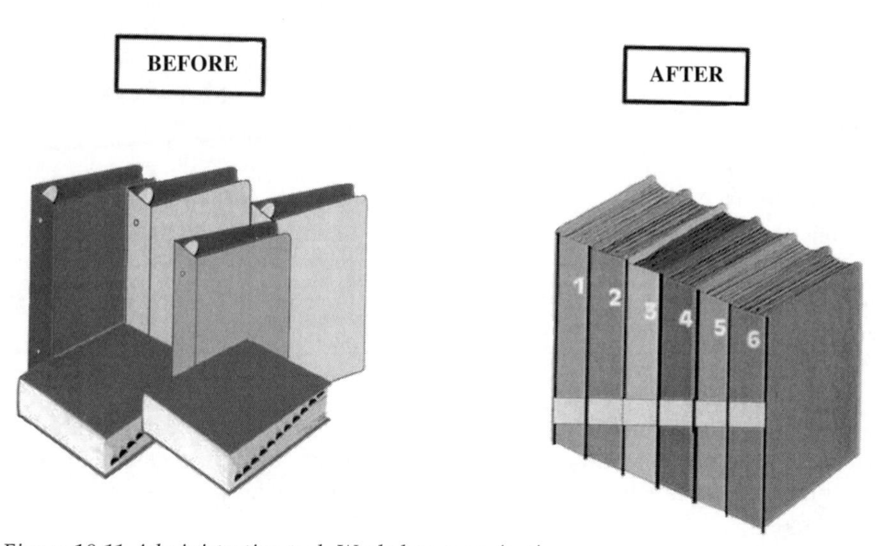

Figure 10.11 Administration tool: Workplace organization

During this session, identify filing/information retrieval priorities. The goal of any 'world-class' information retrieval system, whether manually or electronically filed, is to be able to find what you want within 30 seconds.

The model programme set out in Table 10.3 allows a couple of weeks to consolidate this initial activity.

Sessions 3 and 4

It is useful to add detail to the overview of office systems used to define filing, numbering and labelling processes. Map out the administration process

Table 10.3 Pilot timing plan and supporting notes

Activity	*From*	*To*	*Notes*
Scoping study	1	2	
Mobilization	2	2	Agree who should be involved, communications cascade, etc.
General awareness	3	4	All parties who are involved on administration processes under review, including internal customers and suppliers where possible
Initial training	3	4	Core team, key contact and pillar champion briefing
CAN DO implementation	3	17	Provide the infrastructure to hold the gains
Activity sessions	3	17	Including feedback and presentation of forward plan
Customer needs definition	1	3	(Includes business goals). Ideally during scoping study, but at least prior to team launch
Audit/coaching framework	10	16	Agree with team level 1–4 checklists to their vision
Policy development	8	17	
Roll-out plan development	12	17	Based on roll-out concept agreed during scoping study
Steering Group review	8	17	Monthly from around week 8 when initial policy issues can be reviewed

using examples of documentation and stages of completion. The model should also indicate volumes and areas of loss. If this can be put up and left on a suitable wall, it can be updated with good/bad examples as the project progresses.

Figure 10.12 shows the 'brown paper process mapping' tool. It is a very visual and useful exercise which also helps in teambuilding as well as improving the efficiency and effectiveness of the actual process or system under review.

The stages of the review can include a critical assessment of the process using the standard manufacturing form. The definitions should be amended to use the phrase 'If this part of the system is defective, what will be the impact on . . . ?'.

The elements under consideration should include filing, archiving, labelling and cross-referencing systems. This should include electronically held data. Once the extent and criticality of the system under review are understood, this is followed by a *system condition appraisal*. This activity ensures that the

BROWN PAPER PROCESS MAPPING

- ⊃ To visually map out the A to Z of any process/system/routine
- ⊃ Carry out criticality assessment and condition appraisal
- ⊃ To act as a focus route map for discussion/agreement to eliminate waste by: A B C Z
  - Avoid *duplications* (overlap/excess copies) Duplicate ×
  - Identifying *dependencies* (core and indirect) Dependencies ✓
  - *Defining* inputs/outputs (Links to other systems procedures, measures of performance) →☐→
  - Identifying/*gaps omissions* ? !

⊃ Leading to defining *best practice routines* that can be *standardized* as *maximum value adding support*

Figure 10.12 Condition cycle tool

team have a good understanding of what each element of the system contributes before considering changes (restore before improve).

It is important to agree what is satisfactory as well as what needs attention. From this a restoration/action plan can be developed. Again, these use similar formats to the manufacturing forms.

Where system restoration is necessary, it is important for the team to identify how to prevent the deterioration in the future and how to spot potential problems early. These activities can be loosely grouped under the asset care headings as 'system care' and incorporated as part of best practice.

At this stage, it may be necessary to involve suppliers as key contacts, particularly where equipment is involved or the supply includes on-site activities.

This can be followed up with the development of single-point lessons as part of the definition of core competences and training needs. Once tasks are formalized, the team should consider how to refine the process and allocate the tasks to reduce waste. Consider how to eliminate, combine and simplify activities across all stages of the process, including those areas external to the department. Again, the 'brown paper process mapping' tool can help here.

Sessions 5 and 6

Examples of improvements as part of the problem prevention cycle are shown in Figure 10.13 and illustrate that these are the things we do to hold the gains in the physical and cultural sense (where the 'D' is for the (self) discipline of CAN DO and includes team-based problem solving and prevention techniques like ("5 whys").

It is often possible to assess more than one part of the system. The pilot should attempt to address at least 20 per cent of the most critical.

Once the office systems are restored and standardized, attention can be turned to *process re-engineering*.

Basic principles to be considered include the following:

- Organize around outcomes, not tasks.
- Link activities on a group/parallel rather than individual/sequential basis so that workloads can be balanced and those tasks requiring technical judgement can be isolated and subject to review/simplification.
- Those who use the outputs should perform the process.
- The output from data-gathering activity should provide information in the form required.
- Consider information generation, analysis and use as if it occurs in the same office (don't split tasks by virtue of their location).
- Place responsibility for decisions close to where the work is performed and build control into the process to spot problems early.
- Capture information only once and at the source.

A frequent outcome is for administrative tasks to be delegated to the originator's department in a way which minimizes effort and improves the quality of the information required.

HOLDING THE GAINS: DISCIPLINE

⊃ Before and after photos (Black museum v Standards)

⊃ Colour coding

⊃ Labelling

⊃ Foam cut-outs

⊃ Single-point lessons/instructions

⊃ Audit: To set the standards and pinpoint actions

⊃ Problem solving: Ask why 5 times

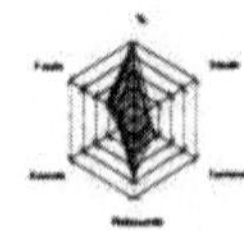

⊃ Post the radar chart results of TPA activity board

Figure 10.13 Problem prevention tools

The workflow analysis is carried out in parallel with the CAN DO steps of define/refine and standardize. This provides a route to support implementation and standardization of working practices as well.

Improvements are introduced on a pilot/trial basis before full implementation to provide the opportunity to refine the best practice based on experience.

This process aims to progressively systemize routine administration through the steps of

- Easy to do right
- Difficult to do wrong
- Impossible to do wrong (zero defects)

Policy development

The issues raised by the pilot should be addressed as they become visible. By using the improvement plan, the type of problems encountered will be in the sequence:

- What to measure and how
- System condition, the level of restoration needed and the type of maintenance activities required to retain system condition
- How to establish and refine best practice
- How to set and deliver improvement priorities
- How to hold the gains and pass on lessons learned

This experience will provide the basis for a realistic and achievable roll-out plan. It will also provide examples to convince others and aid them through the learning curve.

The sequence of the roll-out cascade should be based on the potential impact on business goals. It may also be a precursor to the selection and implementation of computer-based systems.

In parallel with the team activities, the TPA champion and Steering Group committee should develop a policy framework to address the issues raised by the pilot team activity. This may involve redefining roles and the information provided to/from the area under review.

The policy framework should include a progressive audit/coaching process to be applied to all administration areas to deliver:

- Basic systems (measurement cycle and basic information), including CAN DO Step 1
- Transfer of basic lessons (system restoration and improvement), including CAN DO Steps 2 and 3
- Refine best practice and standardize (routine identification and elimination of hidden losses), including CAN DO Steps 4 and 5.

The TPA champion should work with the team to set policy guidelines concerning:

- prioritizing losses;
- definition of routine (team-based) core tasks, early loss detection and routeing of tasks requiring specialist attention;
- definition of technical tasks requiring specialist input;
- definition of core competences, training infrastructure and team recognition processes;
- the sourcing and application of technology.

Table 10.4 TPA core team activity programme

Week		*Contents*	*Activity prior to next session*
3	Initial team launch	TPM principles, review measurement cycle (define APQ loss categories, data capture requirements), establish customer needs and key performance indicators. Define current activity list, and carry out clear and clean planning.	Record current losses, complete first-level activity analysis. Allocation of CAN DO areas. Select location for TPM board.
5	Activity session 1	Identify priority areas from data collection. Carry out outline brown paper exercise to assess filing, numbering and labelling systems. Extend clear and clean to filing systems. Restore obvious labelling/numbering system deficiencies as appropriate. Introduce routine clear and clean activities (e.g. 15 minutes per day).	Focus on CAN DO maintaining the arrangement, preventing unnecessary items from entering the areas.
6	Activity session 2	Audit arrangement progress/status. Identify filing/information retrieval improvement priorities and allocate as part of CAN DO neat and tidy.	Focus on CAN DO neat and tidy retrieval goal, space organization and labelling needs.
7			As above.
8	Activity session 3	Audit neat and tidy progress/status. *Workflow analysis*: Review six loss assessment and select priorities for improvement. Develop detailed brown paper model of priority administration area. Use critical assessment to review each part of the process (include labelling, filing and reference systems). Carry out condition appraisal of critical elements, agree actions to restore system to working condition. Audit neat and tidy progress and allocate CAN DO checking priorities. Focus on reducing the effort required for routine activities such as checking, spotting workload build-ups and workplace/system 'maintenance' activities.	Focus on CAN DO checking and return activities as well as critical system restoration/simplification.

(Contd)

Table 10.4 TPA core team activity programme *(contd.)*

Week		*Contents*	*Activity prior to next session*
9	Activity session 4	Continue with condition appraisal and restoration activities. Define system care requirements (customers and departmental needs) to prevent deterioration and automate update. *Develop best practice* routines for administrative planning, organization and control.	Focus on CAN DO define/refine activities to systemize basic activities and prevent disorder. Continue with system restoration. Trial provisional system care activities and refine best practices.
10			As above
11	Activity session 5	*Improvement activities:* Assess performance against customer needs and vision. Identify improvement opportunities and assess benefits. Concentrate on low cost or no cost solutions. Highlight technical and support problems and identify key contacts as appropriate. Assess/revise roles and responsibilities for routine tasks and those requiring judgement.	Continue with CAN DO define/refine activities. Develop/trial proposals for improved ways of working.
12	Activity session 6	*Make it part of the routine:* Produce single-point lessons and training plans to establish core competences. Identify forward programme. Confirm progress to date and improvement priorities for the roll-out activity.	Carry out skill development programme.
13–15		Additional sessions as required.	
16		Prepare for feedback.	
17	Feedback	For communication and to gain demonstrable commitment from those whose support we need to continue with the improvement process. Present lessons learnt/benefits of TPA. Dry run plus feedback to management with details of progress to date, priorities for action and proposed forward programme.	Mobilization of roll-out planning.

Table 10.5 TPA implementation stages

	Scoping study	*Trial and prove the route*	*Milestone 1 roll-out (everyone involved)*	*Milestone 2 Refine best practice and standardize*
Establish TPA infrastructure Small group activity	Prioritize pilot options, evaluate perceptions, agree pilot programme, infrastructure/roles and mobilize	Awareness and training to develop understanding and skills in the use of TPA techniques. Establish measurement process. Identify roadblocks to progress through the systematic application of the improvement plan. Develop a model of TPA and raise awareness benefits. Add detail to the roll-out concept	Systematically involve all personnel in: Workplace organization Developing written standards for routine activities Improving customer service Identifying hidden losses	Review and refine practices to systemize routine tasks, and trap problems early. Shift from individual to group-based work. Study and improve office tools (paperwork and ledgers). Analyse and reduce duplication, simplify, combine and eliminate tasks where possible.
Organization learning Delivering improved customer service	Define pilot roles and responsibilities, identify awareness and training needs	Review and refine current policy to address issues raised by the pilot team. Establish job core competences	Transfer of lessons learned to other areas. Establish standards for all areas and agreed work schedules to deliver customer service goals	Refine standards and introduce a continuous improvement habit. Reduce lead times and introduce automatic progress co-ordination processes
Top-down Create the environment Set priorities, set expectations, give recognition	Understand and define customer needs. Critically assess the priority of those needs and evaluate current status	Integrate the activities of the core team within the TPM master plan. Co-ordinate the development of audit/coaching framework and links with personal development plans	Establish roll-out priorities. Reinforce standards for workplace organization. Focus on making tasks easy to do right. Encourage a 30-second information retrieval campaign	Encourage a 'one is best' campaign, e.g. one-page memos, one copy filed, etc. Aim to reduce filing space by 50%. Focus on fool-proofing and systemization (difficult to do wrong)

Timing plan

This should take between sixteen and twenty weeks depending on the complexity of the systems under review (see Table 10.4 for approximate timings).

Roll out

Table 10.5 sets out the major implementation stages including the first two roll out milestones.

10.4 Some quotable quotes

In Figure 10.14, we show some 'quotable quotes' from people who have used the TPA improvement tool.

- 'CAN DO 'clear-out' proved very useful'
- 'Improved awareness of how we work'
- 'Identifies scope for improvement'
- 'Means of avoiding replication/duplication'
- 'Gives insight into other support sections'
- 'Good approach for integration of interfaces'
- 'Highlights information gaps/missing links'
- 'Forces us into action'
- 'Speeds up processes'
- 'Makes us challenge and question processes'
- 'Take time out to discuss mutual problems'

Figure 10.14 Quotable quotes from TPA participants

10.5 Why do TPA?

If your OAE is currently 60 per cent, that implies that in any five-day week only three days' equivalent are adding value. The other two days' equivalent is 'a complete waste of time and effort'. If you can eliminate waste and non-value adding activities, so that your OAE is 80 per cent, then that is like having an extra day a week of value adding support. How can you afford not to practise TPA?

11
Case studies

As emphasized earlier, the application of the TPM process varies significantly according to the industry sector and to the particular company or enterprise. The founding principles and pillars of TPM must not, however, be corrupted so that they become unrecognizable.

The key to success is to tailor and adapt the principles to reflect:

- the specific business drivers
- the status and future intentions of other continuous improvement/world-class enabling tools
- the management style and existing culture
- the priority, pace and resource issues implicit in securing a successful TPM journey

One of the best ways to illustrate these vitally important differences is to present case study reports from the different industry sectors and companies concerned.

We are pleased to be able to share the TPM experiences of seven of our customers as follows:

- 3M Newton Aycliffe – Disposable respirators
- Adams (Warner Lambert) – Throat lozenges
- BP Amoco Forties Field – Offshore oil production
- Elkes Biscuits – Biscuit manufacturer
- Henkel – Home improvement products
- RHP Bearings, Blackburn – Bearing housings
- RJB Mining – Coal mining

3M Aycliffe

By Derek Cochrane, Site TPM Facilitator, and Derek Taylor, TPM Co-ordinator

1.0 Background issues

Founded in the United States in 1902 as the Minnesota Mining and Manufacturing Company, 3M now has operations in more than sixty countries, a turnover of $15 billion and employs 73 000 staff manufacturing famous brands such as Post-It Notes, Scotch Guard and Wetordry abrasives.

3M UK employs 4000 people in seventeen locations, with a turnover of £600 million. The 30-acre Aycliffe site in County Durham employs 380 staff in the manufacture and development of many types of disposable respirators for sale around the world.

The 3M group is committed to a stringent environmental and ethical policy and could be seen as having anticipated the empowering culture of Total Productive Manufacturing (TPM) by its long-standing commitment to staff development. Former 3M President, William McKnight, said in his Philosophy of Management paper in 1941:

> Mistakes will be made, but if a person is essentially right, the mistakes he or she makes are not as serious in the long run as the mistakes management will make if it is dictatorial and undertakes to tell those under its authority exactly how they must do their job. Management that is destructively critical when mistakes are made kills initiative, and it is essential that we have people with initiative if we are to continue to grow.

2.0 Why TPM?

In 1994, an initiative called 'Training for Success' highlighted seven areas where there were clear opportunities to move forward: teamworking, continuous improvement, systems, people development, communication, training and culture.

TPM provided the mortar to link the seven building blocks of 3M's own programme and they therefore decided to adopt it across the site in 1995. It is now the umbrella under which every other initiative hangs, embracing the philosophy of continuous improvement.

3.0 Benefits

Overall

TPM at 3M Aycliffe is spearheaded by two members of staff – Site TPM Facilitator and Engineering Specialist, Derek Cochrane, and TPM Co-ordinator, Derek Taylor.

Derek Cochrane sums up what TPM means for 3M: 'TPM involves everyone at all times, working together to improve their equipment, processes and methods and produce quality products that will surpass our customers' expectations. It promotes teamworking, training and people development through its systematic approach of involvement and ownership.'

After originally working on three pilot areas, there are now four main areas of the site, each managed by a dayshift group leader. Each area consists of improvement zones and TPM core teams run by shift team leaders. Nearly everyone on site has now had TPM general awareness training.

Derek Taylor comments: 'At each stage of the journey, WCS worked closely with teams, managers and facilitators, the union and key contacts to promote in-house ownership.'

Site-wide improvements made under TPM

Various improvements were made, including net annual cost savings for:

- 42 proactive TPM teams
- TPM team activity boards
- Each piece of equipment is labelled with the name of its team leader and corresponding maintainer
- Asset care routines for operators and maintainers on all equipment
- Introduction of single-point lessons (SPLs) to offer a concise and easily accessible training guide for processes or pieces of equipment
- Computer systems that log unplanned events direct from the machine computers to a central database
- Training facility built and in-house training courses developed
- CAN DO – Cleaning, Arrangement, Neatness, Discipline and Order – adopted on the shopfloor, in the stores and in administration to provide a world-class workplace as in 5Ss
- Visual indicators on machines, for example, colour-coded dials marked to show safe operating limits
- Shadow board for tools
- Disciplined quarterly TPM audit system with milestones and levels
- Operators and maintainers involved in early design reviews before acquiring new equipment (TPM for Design)
- A TPM team was introduced in the maintenance stores which completely revolutionized the stores system, ensuring that spare parts are always readily available and providing foolproof systems for locating and logging out items

People benefits

- People development
- Enhanced skills
- Greater staff involvement
- Improved morale – operators' ideas are implemented with an acknowledgement that engineers don't always have the only or best answers
- Improved trust and a more open relationship between operators and maintainers and between teams working across different shifts
- New identification of training needs through analysis of what people do know versus what they could or should know – opportunity to increase skills and knowledge
- Improved communications
- A more motivated workforce
- A strong sense of ownership and pride in the workplace

Business benefits

- Improved OEE, thus releasing the inherent productive capacity
- Improved working conditions
- Machine problems addressed instead of lived with, leading to increased efficiency
- Improved quality – product and processes
- More effective return on capital employed
- Greater reliability of machinery, and so less reactive and expensive maintenance
- Better able to meet and exceed customer expectations
- Able to predict areas of wear in machinery
- Safer working environment
- Prevention of breakdowns
- Prolonged life cycle of equipment

4.0 Continuous improvement and TPM

Overall

3M Aycliffe uses TPM as the means of providing an opportunity to all staff to contribute to the programme of continuous improvement.

Continuous improvement for 3M Aycliffe also means continuous assessment. A stringent TPM audit system is in place whereby each improvement zone is assessed according to two milestones under which sit four levels of achievement. Each improvement zone is identified with a plaque clearly showing the level it has attained. In order to move through the four levels, each team has to hold gains of the last level and will, therefore, be reassessed on previous levels.

Bottom-up and top-down audits are completed quarterly and the results are then compared to pre-set benchmarks. The Aycliffe site also runs an annual recognition scheme, with targets set for OEE results and team level status.

TPM in the stores

The stores at 3M Aycliffe, which holds accessories and spare parts for all machines, also underwent a TPM-driven overhaul two years ago. The changes that were made as a result have completely revolutionized the stores system, ensuring that spare parts are always readily available and providing foolproof systems for locating and logging out items.

The stores now has a precisely ordered shelving system by part manufacturer, a database search facility and single-point lessons for locating items, how to book out items, and returning items for repairs.

The classic six losses have been modified to apply to the stores system, to enable a stores OEE to be generated, so that efficiency can better be assessed.

The maintainers are now responsible for ensuring that all the parts they believe they require are available in the stores, and this is audited as part of their team's quarterly review.

Quotable quotes

The attitude of staff towards TPM underpins its success:

> I classify TPM as a major success story for the Aycliffe site. It has moved from the status of Project to that of a Continuous Improvement Strategy which incorporates the Total Engagement of our people in its implementation. The activity surrounding TPM is now truly organic, laying down fundamental principles of working methods which the plant will build upon for the next Millennium.
>
> **Dave McDonald, European Manufacturing Manager**

> Before TPM, we just knew what we needed to do. Now we have a more in-depth knowledge of the machine and a greater understanding. We didn't like TPM at first, but now we've really got into it.
>
> **Val Cowper, Operator**

> TPM is not a cleaning exercise. It is a strategy which is process and results oriented. It provides simple methods for data collection, analyses, problem solving and process control. This has allowed us at Aycliffe to move the continuous improvement ethic forward, significantly, to the benefit of both the business and everyone employed here. TPM is not about technology – it's about people.
>
> **Tony Hall, Production Manager**

> Through TPM, the teams are now encouraged to be self-motivated, challenge existing practices and develop new methods to improve. Due to the improved communication and a greater sense of team responsibility, we now have a unified area.
>
> **Leslie Naisbitt, TPM Team Leader**

5.0 The future for TPM

3M was celebrating recently after its CAF (Combined Assembler Finisher) team won the prize at the International TPM Conference, TPM6, held in Stratford upon Avon in June 1999.

Awarded to the UK company which has made the most outstanding progress in its implementation of TPM, the winning team's projected annual improvements, applied across nine CAF machines, totalled over £300 000.

Net annual cost savings included:

- Changing sensors and modifying the shell cutting method to prevent problems when welding the perimeter of the face mask.
  Saving: £18 300 per machine
- Valve unit changed and adjustments made to fingers and vacuum to prevent excessive glue contamination on noseclips.
  Saving: £5400 per machine
- New system implemented to check Modicon units upon receipt from repairers as they persisted in breaking down.
  Saving: £2300 per machine

The award is a testament to the achievements made by the company to date in its implementation of TPM.

3M will continue to coach the TPM teams through the milestones and levels to ensure that 3M Aycliffe is a world-class manufacturer.

In the new Millennium, the site is striving to link its continuous improvement activities into the business holistically and will be investigating ways of benchmarking progress.

TPM is about continuous improvement. If you improve, you get better and become more competitive. The process of TPM is never-ending, a journey with no final destination. For 3M, the future will see TPM become a culture and not just an initiative.

Adams (Warner Lambert)

By Clive Marsden, Technical Director, and Chris Rose, TPM Manager

1.0 Background issues

Famous for producing the world's biggest-selling medicated sweets, Halls throat lozenges, Adams (Warner Lambert) is a company synonymous with quality pharmaceutical, healthcare, shaving and confectionery products. Listed among the world's top 100 companies, Adams counts Wilkinson Sword, Benylin, Clorets, Dentyne, Stimorol gum and Tetra among its famous brands.

In 1994, the then Maintenance Manager, Clive Marsden, commissioned an audit of the maintenance function at its Manchester factory, which employs 500 staff and produces approximately 600 tonnes of sweets a week. The results highlighted a shortfall in data, a lack of operator involvement in the maintenance function and organizational issues. As a result, Clive developed a plan and gained enthusiastic backing – including financial backing – from Alan Hulme, Plant Director, along with the Senior Management Team, to address some of the issues highlighted by the audit.

As a result, a Computerized Maintenance Management System (CMMS) called Mainsaver was proposed, which, among other things, would provide data to improvement teams. In addition, a wide-ranging Total Productive Maintenance programme was planned, complemented by a restructuring programme to enhance teamworking.

This was seen as a major change to a very traditional site, and protracted negotiations with both unions took place before the plans were implemented.

It was very interesting to note that once the unions – particularly the engineers – were 'on board' with the proposals, they became fervent champions of – and heavily personally involved in – TPM.

2.0 Why TPM?

Adams was looking for a shopfloor-based, easily understood continuous improvement tool. TPM was the perfect solution in that it provided a grass-roots process, with a simple and universal business performance measure of OEE (Overall Equipment Effectiveness), and involved both operators and maintainers as joint experts and teamworkers.

TPM was formally introduced to the Manchester site in 1996 – the first Adams site in the world to implement TPM – with three pilot projects, each focusing on a critical piece of machinery or group of similar machines: Volpak, 1001 and 2001 machines.

Leading TPM consultancy, WCS International, provided initial planning and training expertise, to develop a long-term sustainable TPM programme that is now supported by coaching and regular audit and review advice.

Based upon the successes of the pilot teams, it was decided to 'roll out' the TPM process across the whole site. Clive Marsden gained agreement to recruit a full-time manager to co-ordinate and lead this 'roll-out'. As a result, Chris Rose was appointed as full-time TPM Manager two years ago, having worked at Adams for 22 years, where he began as a maintenance craftsman.

Full-time facilitator

Chris Rose co-ordinates TPM activities at the Manchester site. He also facilitates the TPM process, attends team TPM meetings and hosts his own TPM Facilitator's meeting once a week. Every four weeks, Chris reports back to the Area Steering Group, which addresses any roadblocks and ensures that the team's activities are linked to the company's business goals.

Communication is the key

One of the most important aspects of the TPM programme is the communication of its results and initiatives – there was a time at Adams when TPM was thought to be an elitist hobby for a select group of workers. However, this issue has been addressed through the creation of improvement zones and involvement of staff through the company Intranet site.

Wrapping Technical Operator and computer enthusiast, Fiore Festa, developed a TPM site for the company's Intranet system in his spare time. Chris Rose was so impressed with Fiore's work that he showed the Plant Director what he had done, and Fiore was seconded from the shopfloor so that he could develop the project.

Workers from all departments can log in and take a look at the work of the different TPM teams. Photographs of each piece of machinery can be found on the relevant TPM team pages. By clicking onto a specific part of the machine, a new screen is revealed which gives details of TPM actions and cost savings for that area of the machine.

Chris Rose comments: 'The TPM Team page on the Intranet site has been invaluable in educating all staff about what we are doing with TPM.'

The site can also be accessed by Adams factories and offices overseas.

3.0 Benefits

For all staff at Adams Manchester, TPM is a common language that previously did not exist – and OEE is the cornerstone of this language. Says Chris: 'Everyone, from finance to operators and maintainers, know what the OEE is and the significance of improvements.'

A recent Human Resources (HR) audit carried out by an American team established that the introduction of TPM had improved company worker morale, as people felt they were being involved in the decision-making process.

Involving finance

The Finance Department has been heavily involved in the implementation of TPM. Manufacturing Accounting Manager, Liz Morton, comments: 'We had to be able to show that real cost savings were being made with TPM or why do it? TPM cost money in that we were taking staff from the shopfloor to engage in TPM meetings and training. We had to ensure that this was going to be money wisely invested.'

Liz is also involved in the production of annual business plans. TPM, and the costs and savings involved, are a vital part of this plan.

Liz Morton spent time with the TPM teams and calculated a value for a 1 per cent improvement in OEE. Continues Liz: 'Putting a value to OEE improvements was a good starting point. It provided a way to bridge the gap between TPM and the shopfloor, and management and finance.'

In terms of value for money, one key area that Liz looked at first was process bottlenecks. Explains Liz: 'At Adams, individual machines do not operate in isolation. This means that unless you first target bottlenecks in the process, you could waste a lot of time making improvements that never show up on the bottom line.'

Through her role in finance and a focus on activity-based costing, Liz Morton has assessed the big picture and agrees that TPM improvements can have an enormously beneficial knock-on effect. She says: 'If a machine breaks down, then we have no product being produced, staff being paid to do nothing, plus the cost of repairs in terms of labour and spares. If you then look at the fact that this product may be part of an urgent shipment, there are implications for customer service. If we want to meet the customer's deadline, then we have to look at getting staff in at the weekend, plus using air instead of sea freight. The costs just snowball.'

Liz also focuses on cost deployment, which highlights the cost of not making things 'right first time' (cost of quality). TPM's six losses were top of the hit list. The benefits of cost and quality deployment are:

- As a motivator for teams
- Teams are able to make financial decisions (empowerment)
- TPM teams can speak senior management language (i.e. £s)
- Focuses on costs

It's the people that matter

TPM is often associated with people empowerment, and Adams has working examples of this. Chris Rose continues: "We have found that TPM teams are responding directly to business requirements in terms of production – for example, in finding ways to meet customer requests for amended packaging. The team will look at trialling a solution or new way of working and then report in to management to request the financing required to fully roll out the improvements. Previously, capital expenditure requests were the sole remit of management."

The environment in which TPM thrives, along with the workers, is one of flexibility and versatility. Jobs are developed away from previously rigid roles. Operators, as opposed to electricians, now calibrate some machines. This frees up the electricians to work at a more technical level, which in turn enables proactive, instead of reactive, working to take place.

TPM teams also come up with best practice guidelines for each piece of machinery. These are incorporated into training manuals and videos and disseminated across all shifts.

TPM for Design principles are also applied to new machinery, through involving the experience of the operators and maintainers, with minimum OEE levels negotiated into the terms of purchase and commissioning.

TPM has benefited safety and environmental issues in a number of ways. An element of the criticality assessment covers safety, and any improvement in looking after the equipment will have knock-on benefits in this area.

Not least, it has helped raise awareness with regard to safety issues and empowered those same workers to effect a cure or remedy with the help of key contacts such as the Health and Safety Officer.

Benefits

- Work is made easier by working smarter, not harder.
- Better understanding of engineering and production problems.
- Operators have a more in-depth understanding of the machine.

- There is more cost awareness, not only of improvements in OEE, but also of material costs. Workers are therefore more conscientious and focused about waste.

Quotable quote

> TPM has enabled people to develop skills that they didn't know they had.
>
> **Chris Rose**

4.0 Team focus

Volpak bagging machine

The Volpak machine packs 120 bags of sweets every minute and was the focus of one of Adams' three pilot studies. The machine was part of the 'New technology' introduced into the factory over the past few years and had been giving problems with quality and availability, partly due to the new type of bag it was handling. Through collecting OEE data, assessing the six losses and setting improvement priorities, dramatic improvements in performance were made.

A 1 per cent improvement in performance on the Volpak is equivalent to £35 000 in annual cost savings. In just three months, a series of improvements resulted in the OEE leaping from 69 per cent to 78 per cent, giving £210 000 of cost avoidance and £105 000 of actual savings per annum based on an original estimated and planned OEE for the *new* machine of 75 per cent. Finance was then able to incorporate a target OEE improvement of 80 per cent, giving a further saving of £70 000, into the budget for the following year.

GD wrapping machine

When a reel of foil sweet wrappers ran out, it took approximately 4 minutes to change the reel, with a loss of 50 wrappers and an average of 22 reel changes per shift.

The TPM team responsible for the machine looked at an electronic heat-impulse foil splicer, an improvement which led to £16 220 in cost savings for the factory.

5.0 The future for TPM

The future objectives for TPM are to:

1. Reduce material usage.
2. Integrate TPM fully into the Adams culture.

3 Start new teams and integrate team leaders into the existing TPM teams.
4 Use TPM to support the maintenance strategy.
5 Promote interdepartmental communication by staff from one department joining a TPM team that supplies them with a service.

At Adams, TPM is now a way of life – both on the shopfloor and in the boardroom. From the successes with TPM at Adams Manchester, it is easy to see how, once begun, there should be no reason for going back.

BP Amoco Forties Field

By Warren Burgess and Mike Milne,
BP Amoco Operations Excellence Facilitators

1.0 Background

In 1997, BP's Forties Delta platform pioneered a series of Continuous Improvement (CI) projects that have helped reduce unplanned shutdowns by 53 per cent and set the stage for the future operations of the Forties platforms.

The various projects, managed in conjunction with WCS International, have pulled together platform and beach workers from every discipline in a united cause: to improve the safety, environmental impact, efficiency and productivity of BP's operations on the Forties Delta Platform.

The methodology used borrows heavily from Total Productive Maintenance and has been introduced as a CI improvement tool, along with other improvement initiatives in safety (STOP) and training.

Rather than starting 'yet another initiative', this complements the concept of building on existing good practices as a practical application of organizational learning and personal development, with the goal as Totally Productive Operations (TPO).

Initial workshop

The work began in March 1997 with a short planning exercise followed by a four-day 'hands-on' workshop. This looked at two pieces of equipment to provide awareness and training, and a pragmatic demonstration of the effectiveness of the approach.

The two projects undertaken during this workshop were the sodium hypochlorite and scale inhibitor systems. Even though the four days were mainly an opportunity for delegates to experience the power of CI, the two teams identified many benefits.

The scale inhibitor team developed recommendations to reduce maintenance intervention and the sodium hypochlorite team proposed an alternative design of pump which significantly reduced the estimated capital expenditure investment already allocated to this system.

Project focus

The initial workshop projects were followed by focused pilot projects to improve the maintenance and operability of four key areas of plant:

- MOL pumps
- Seawater injection
- Hydrocyclones
- Separators

These were chosen by representatives of the complete platform team at the Platform Conference. A third of each shift team dedicated itself to improvements in one of these four areas, using a nine-step TPM-based improvement plan.

Initial difficulties

Continuous Improvement Facilitator, Mike Milne, sums up the problems encountered at the start of the project:

> We had many difficulties to begin with. First we had to address the problem of getting the time to attend the CI meetings. Teams were tied to a spiral of reactive performance that did not leave enough time for proactive measures.
> Platform management stepped in to set priorities that ensured we broke free of this restriction: Safety, Production and CI and the promotion of a greater awareness of the benefits of the proactive approach.

Platform personnel also attended the 'Manufacturing Game' – a business simulation, developed by Dupont, that drives home the benefits of running a proactive business. Back on the platform, the teams planned and organized a two-hour slot to meet and discuss their CI project. Shift Team Leaders (STLs) needed to provide resources that freed staff to attend the meetings. As part of the Trip Objective process, measures were put in place to ensure that this happened. 'This was not an easy target and it challenged the planning skills and resourcefulness of each STL and CI team leader,' concludes Mike.

2.0 Results

As we move into 1998, the projects are currently in their final stages and have already brought a variety of specific and measurable benefits to the successful operation of the platform – the most notable being the reduction of unplanned shutdowns (see below).

Introduction of procedures

A series of standardization procedures have been introduced, including checklists, best practice routines, asset care procedures and problem-solving techniques, which have increased the Overall Equipment Effectiveness (OEE) in all four pilot project areas.

STL, Gerry Scanlan, observes:

> The project has given the team a lot of personal satisfaction and a great deal of hard work went into it. Many Best Practices have been developed which have been rolled out to the other two shifts that are working back to back with ours.
> This cascading of information from shift to shift, and ultimately having the agreement of all four shifts, is vital to the success of the CI Initiative.

MOL pumps

PO6 is one of the three turbine-driven pumps used on the Main Oil Line (MOL) system. These have all become less prone to unplanned shutdowns. PO6's best fit OEE is now at 50 per cent as at end December, up from 6 per cent in May 1997.

Howard Dickson, MOL Team Co-ordinator, comments:

> PO6 was the worst of the three pumps. A Critical Assessment highlighted a pile of smaller jobs that needed to go through the maintenance system. We had to establish a fault-reporting system as a framework to get the jobs done – all have now been actioned or scheduled.
> Some jobs are just too big and so needed onshore support. This opened up communication lines with the engineers and vendors. Everyone now knows in whose area things belong. We also had some very good discussions among the techs themselves – before this hands-on team-based approach, we just hadn't been aware of the problems other disciplines were having.

Howard concludes: 'The structure of the 9-step improvement programme helps you do it the correct way, even though it can seem a bit heavy.'

Seawater injection

Two excellent results were obtained in this area. The OEE has risen from 45 per cent to 90 per cent and the corrosion control performance has risen from 42 per cent to 89 per cent.

Hydrocyclones

The main breakthrough for the hydrocyclone team was in identifying the problem with the inlet quality from the separators. As a result, both Bank A and Bank B hydrocyclone facilities greatly increased their OEE rates – Bank A from 40 per cent to 55 per cent over the year, and Bank B up to 80 per cent from 10 per cent – an excellent result.

Peter Hanson, Hydrocyclone Team Co-ordinator, comments: 'Prior to the project, Maintenance Technicians had no idea of the Operations side and vice versa. It was a real success in terms of the training aspect and brainstorming equipment issues.'

Separators

Again, two significant results from the two vessels involved in the pilot project, with OEE up from 30 per cent to 88 per cent on VO1 and up 60 per cent on VO2.

Unplanned shutdowns reduced by 55 per cent

Perhaps the result with the biggest impact generated by the project is that concerning the reduction of unplanned shutdowns. In 1996, unplanned shutdowns were running at an average of 5.4 per month. The target for 1997 was set at four per month – a 31 per cent reduction year on year. The reality was that unplanned shutdowns averaged just 2.6 per month – a reduction of 55 per cent. This figure greatly affects the rate of oil production and is the main reason why Delta has had a dramatic increase in its oil production in 1997.

Broken down, the figures for the second half of 1997 look even more impressive. During the first half of the year, unplanned shutdowns were running just below the target, at 3.8 per month. However, from July to December there were only nine unplanned shutdowns – 1.5 per month. This really is an impressive statistic. All on Delta onshore and offshore can be proud of this achievement. However, it does need to be sustained and, although there were only a total of 31 unplanned shutdowns in 1997, if we consider external events that caused FD to shut down, there are still considerable improvements to be had in striving towards a goal of zero unplanned events.

This is one of the main themes of continuous improvement and, just to prove it can be done, Delta achieved zero unplanned events in August, September and November 1997.

Looking specifically at some of the CI project areas, the MOL pumps achieved a remarkable improvement in unplanned shutdowns, down from thirty-seven in 1996 to nineteen in 1997. Again, the second half of the year accounted for only four of these unplanned events. The other significant improvement was in the area of Natural Gas Lift (NGL) unplanned shutdowns – these were reduced by 75 per cent in 1997.

ECS rates as low as 5 mg/kg

One of the Forties Delta's key environmental targets is to develop and deliver options to reduce environmental emissions. The legal limit of oil in water discharged to sea is 40 mg/kg. However, Forties Delta set an environmental target of 23 mg/kg at the beginning of the year.

Through the work of the separator and hydrocyclone teams, Effluent Control

System (ECS) rates of between 5 and 10 mg/kg have been achieved. These results will need to be stabilized and achieved consistently throughout 1998, but nonetheless represent an excellent starting point, giving the platform great confidence in its ability to reduce the environmental impact.

3.0 Future focus

Having seen that their hands-on involvement is key to securing improvements on the platform, the separator CI team were eager to become involved in all phases of the upgrade works planned for the test separator, VO3, from concept through design to commissioning. Other members of the integrated operations team came forward to offer help to set the criteria for the new equipment.

For Offshore Installation Manager (OIM), Brian Barnes, one of the key benefits of CI initiatives throughout 1997 has been the continuing evidence of people involvement and motivation. Comments Brian: 'Normally it would just have been the engineering teams onshore who were involved in the specification and procurement of new equipment. This time, their work has been given greater focus by the enthusiastic exchange of ideas and information from offshore operation team members.'

Previously onshore engineering teams would have approached the Operations Team Leader (OTL) for information on new equipment. It is now the operations team members themselves who are approached for feedback and information on projects like VO3 upgrade.

Improvement zones

For the next stage of the CI programme, the platform was divided into eight geographical improvement zones. The principles learnt during the four pilot projects of 1997 were applied to each piece of critical equipment in the various improvement zones.

This time around, every member of the shift team will be involved, highlighting the need for effective dissemination of information. Again, they will be encouraged to work alongside the onshore engineers to anticipate and manage changes to machinery and plant equipment.

Comments Brian: 'The shift in the way people worked together on VO3 led naturally to the new way of co-operative working that will certainly help to drive the improvement zone approach – a key focus for the CI work for 1998.'

Facilitator, Mike Milne, adds: 'Ultimately these steps ensure that the customer gets what he wants – where the customer, in this case, is the operator of the equipment.'

Early equipment management

The work on the improvement zones will be a crucial part of introducing the concept of Early Equipment Management, i.e. ensuring that all aspects of operations and maintenance are addressed at the design stage for new equipment modifications by the operations and maintenance people. In 1998, Forties Delta saw the installation of several pieces of new equipment, both capital investment and operation and maintenance works. It was, therefore, vital that the team members were fully up to speed with all the skills learnt so far in the CI programme, in order to contribute effectively to this next level of proactive involvement in the development of their facilities.

Brian Barnes and Mike Milne identified direct links between the improvement zones on the platform and work required for these equipment upgrades. Responsibility is being allocated to on- and offshore staff for each main activity – with both sides working closely together. For example, an offshore focus group has been established so that onshore engineers can discuss equipment issues, and shift teams are, therefore, already becoming familiar with the crucial role they will play during the installation phase, and its implications for the future efficiency of the platform.

And, of course, as is the case with all of the CI work done so far, this new approach will give the operators a major say upfront in how new and existing equipment is to be run and developed.

Training focus

Forties Delta has also embarked on the 'STOP ... for Safety' programme, and platform staff are currently being trained in STOP techniques. The programme is backed up by the Advanced Safety Auditing process, which uses similar techniques to those of STOP. Again, this is a CI process, with team members and management learning new ways to improve Forties Delta's already impressive safety performance.

Team members with a supervisory responsibility are continuing their training programme, and by the end of 1998 every member of platform staff will be trained in STOP . . . for Safety techniques.

Cross-training

One of the most important focuses for training in 1997 was cross-discipline training, where operators from every shift have been taught not only to focus on their own specific field of expertise, but to learn about the wider application and impact of their role by training in neighbouring areas.

The learning of new skills, while retaining their core discipline, has allowed members to utilize CI more fully, producing a greater understanding of the parameters that drive the plant efficiently. For the individual, this spread of knowledge means an increase in responsibility, awareness, effectiveness and recognition.

Mike explains: 'We now have an Integrated Operations Team with a flexible approach to operating and maintaining the plant.' Mechanical, electrical and instrument technicians are capable of carrying out duties that were previously carried out only by a production technician. Meanwhile, production technicians assist in maintenance tasks and other asset care routines previously the domain of the maintenance technician.

Comments Brian: 'If the operators have a better feel for how maintenance links into the whole production process, they have far better judgement when it comes to making on the spot decisions that can drastically affect safety, performance and efficiency. We now have Instrumentation staff who can run the NGL plant and Production Technicians training in Control Room duties – a year ago that would never have happened.'

A vital element of cross-training is the positive attitude that it engenders – an openness and willingness to exchange knowledge with workers from other disciplines, giving a real team spirit right across the platform and onto the beach.

Ultimately, the drive for CI means that the plant is more stable. People are released from a round of reactive responses to machinery problems and instead can take a step back to implement proactive measures that allow both the individual and total platform team to be more in control of its operations: prevention rather than cure.

4.0 The future for Forties

The importance of the work on Delta cannot be underestimated. In the future, Alpha, Bravo and Charlie will all share in the CI practices tried and tested during the late 1990s by Delta. The intent for the Forties Field is to move to consistent use of best practices across the whole field and maximize the opportunity of sharing lessons learned between each of the platforms. Other assets outside Forties are already taking an interest in the work that Delta have embarked upon and are keen to learn from the Delta experience.

Over the next five years, Forties Field will also be looking at a series of environmental improvement issues, including the emission of hydrocarbon gas, carbon dioxide and overboard discharges.

So just as an operator now shares knowledge with and co-operates with a team member from another discipline, so too the Forties Platforms will pool their strengths and learn best practice from each other to everyone's benefit.

Already Delta staff have been involved in seeking best practices and new ideas from other assets, both at home and abroad. This will continue throughout 1998, with several visits being planned. Platform staff are about to embark upon a shared learning experience. The programme will include short-term transfers of personnel. Delta have already been hosts to other BP asset staff. They believe that by this type of co-operation they have a forum for exchanging techniques, methodology, procedures and processes that has been shown to improve business results.

As BP Chief Executive, John Browne, said in a recent *Sunday Telegraph* article: 'The most important thing I have done is to play a part in building a team for today and for the future. We set the strong goals and within that our people innovate to get the right answer.'

Elkes Biscuits

By Ian Barraclough, TPM Manager

1.0 Background

Founded in 1924, Elkes Biscuits began life as a small tea and cake shop in High Street, Uttoxeter. The company now has a turnover of £55 million, manufactures 1000 tonnes of biscuits per week and employs 1300 people.

Owned by Northern Foods, Elkes Biscuits is primarily a private-label supplier. Included in the product range are well-known biscuits such as Custard Creams, Nice, Farleys Rusks, Ginger Nuts and the most famous of them all, Malted Milk, which were first created over sixty years ago.

2.0 Why TPM?

In a commodity product like biscuits, driving down costs is a continuous process. One way of reducing costs is to minimize downtime, which is where TPM excels. Following a visit to the 'TPM 4' Conference, Elkes Biscuits has introduced a variety of significant and ongoing improvements to the plant. These centre around:

- new staff structure, which has merged front-line maintenance and production under operations and removed a reporting layer;
- full-time TPM Manager, Ian Barraclough;
- asset care and best practice routines for seventeen projects to date, including high quality and highly visual single-point lessons;
- introduction of operator technicians, allowing skilled maintenance technicians to focus on more proactive, advanced jobs;
- move towards multi-skilling, application of asset care and best practice routines by operator technicians;
- dedicated TPM Centre with training room, computers, manuals and library;
- nine-step TPM process for critical pieces of machinery;
- gradual training of the whole workforce on the TPM nine steps;
- high-profile activity boards showing in detail the nine steps for each machine, including spare parts log, OEE performance bulletins, CAN DO audits and TPM updates;
- daily review system of line performance using OEE data;
- fortnightly continuous improvement group meetings to review and plan activities.

3.0 Benefits

One of TPM's strengths is its ability to operate alongside other quality initiatives. At Elkes, these include NVQ assessments, the drive for Investors in People accreditation and FAST (Elkes' own Faster Achievable Set-up Times – similar to SMED).

TPM, working with these other initiatives, has helped to secure the following benefits at Elkes:

- engineering, maintenance and production working closer together to resolve problems;
- TPM used as a driver for problem solving *and* problem prevention;
- more effective teamworking within and between departments;
- improved quality and performance of machinery;
- financial benefits through production line OEE improvements of 5 per cent to 10 per cent, as a result of improved machine performance.

A small example of exactly how TPM affects life on the shopfloor: the in-line creamer had a rather antiquated and inaccurate method of aerating the cream in the biscuit sandwich. As a result of the application of TPM principles, the system was changed, resulting in finite control of the process, improving the biscuit quality and consistency, and making the operator's life easier.

Before TPM we would identify problems but had no chance to do anything about them. TPM is great because it gives us the time and opportunity to make changes which last because we underpin the problem with effective future asset care, best practice and single-point training.

The nine-step process gives a very structured and cohesive approach to TPM which is both simple and thorough.

4.0 The future for TPM

Elkes takes TPM very seriously. It is currently used on the majority of lines at the plant and there is a firm commitment to continue to roll it out plant-wide over the next couple of years.

A key project for the immediate future will be the application of TPM to the latest creams plant investment at Elkes where every piece of equipment in the facility will be brand new, providing the perfect platform from which to launch TPM.

Henkel

By Gordon Hill, TPM Facilitator

1.0 Background issues

Henkel Consumer Adhesives is famous for producing the world's biggest-selling adhesive brands, including Unibond, Pritt, Loctite, Solvite and Copydex. The site in Winsford, Cheshire, manufactures 28 000 tonnes of home improvement products each year. With over 900 product types, its famous Pritt glue stick has 80 per cent market share and enough Solvite wallpaper paste is bought each year to paste a roll of wallpaper thirty times around the world!

Its parent company, Henkel, employs 55 000 staff in eighty countries and manufactures 11 000 products with an annual turnover of £7 billion.

In 1996, an audit of the maintenance function at the Winsford site was commissioned by Operations Director, Mark Hamlin, and Engineering Manager, Mike Williamson. As a result, a Computerized Maintenance Management System (CMMS) was introduced to manage the data necessary to evaluate production effectiveness. As no off-the-shelf database system had the functionality Henkel was looking for, the company decided to develop its own system using Microsoft Access.

With a background in computer systems, Business Process Manager, Gordon Hill, was seconded into the Engineering Department as full-time TPM Facilitator, with the objectives of managing the introduction of Total Productive Maintenance throughout the plant and developing the CMMS.

The focus of TPM made Henkel re-evaluate the way in which data was collected. They discovered that most of the information was not only already available, but often duplicated in a series of forms which all ended up in different places, some never being used.

One of the first things Gordon Hill did, therefore, was to develop one all-inclusive input form containing all the necessary data required for the CMMS. Each form has a workings section for individual operators to calculate their shift Overall Equipment Effectiveness (OEE), involving them in the process of data collection and processing and therefore giving ownership for the quality of the information.

The database system played a big part in the flow of information. The results of OEE data input into the CMMS are fed back to team leaders each morning. The question is then asked each day: 'Did we achieve the production plan?' But whereas before the answer Yes or No would simply be collected for discussion at a weekly management meeting, this time if the answer is No, the next question will be 'Why? Why? Why?'. If the answer is Yes, then follows, 'At what cost?'

Effective communication between the shopfloor and the maintenance function at Henkel was hampered by an over-complex organization structure. The structure of the management team was therefore reduced from seven layers to four and the Operations Manager united the two functions of production and logistics.

On the shopfloor, production teams were organized by process rather than function. Each team became multi-functional, including members from planning, materials movement, warehousing and, of course, maintenance. The core team was then supported by a number of key contacts from Quality, Engineering, Design, Finance, Purchasing, R&D and Marketing.

Existing supervisors were developed into ten team leaders, where each team leader is an integral part of the production team rather than simply being another layer of management. Each team was then given a deputy to support the team leader and to act as the team's trainer. Together the teams and their key contacts can now focus on the elimination of losses across the whole supply chain.

2.0 Why TPM?

Says Gordon Hill: 'Many initiatives have made attempts at improving productivity, but only one, TPM, provides a pragmatic approach that can be understood by everyone at the grass-roots level: namely a continuous drive to increase OEE.' Henkel found that another of the big strengths of OEE is that it gives the ability to measure the cost of poor quality in monetary terms.

Continues Gordon: 'The OEE is at the heart of TPM. Many companies believe that TPM is limited to being a maintenance department-driven initiative. It can be, but it is much more effective to use TPM as a holistic approach to equipment and operational effectiveness, involving finance, operators, management and even administration staff.'

With the concept of the CMMS established, TPM was launched, via the following stages, from June 1997:

1. A pilot study, supported by leading TPM consultancy, WCS International, helped Henkel to understand the impact and scope of TPM on productivity improvements.
2. This was followed by a number of awareness presentations for all 240 staff members, giving an overview of TPM.
3. WCS then conducted key contact training for a number of middle managers to give them a more in-depth understanding of TPM and to gain their personal input and commitment to the process.
4. Finally, staff directly involved in the pilot project attended a four-day hands-on workshop to experience everything that the teams would cover in the following months. This gave a clear focus for future TPM activity.
5. The pilot was concluded with a feedback presentation at the end of 1997 and, because of the significant and measurable improvements achieved, it was agreed that TPM would go live across the site in July 1998.

A condition appraisal, refurbishment and asset care routine was implemented for each piece of machinery by the end of October 1998, followed by the in-house retraining of all operators. Maintenance fitters meanwhile completed cross-craft training in order to improve the flexibility of the maintenance department by removing 'mechanical' and 'electrical' barriers.

A range of best practice routines were developed by the teams via the introduction of single-point lessons (SPLs) for each machine and will be rolled out across the site in the early part of 1999. The SPLs act as highly visual and simple to understand training aids and are particularly invaluable for Henkel due to seasonal peaks of temporary staff.

The SPLs are used in conjunction with a four-sector competency wheel so that the aptitude of each staff member for each SPL for each machine is monitored and developed in the form of a training matrix. Flexibility is a key factor, so every operator has a training plan that ensures capability and competence across a number of lines.

The SPLs will be launched through the personnel function, and so TPM is becoming a significant part of the Human Resources Department, as well as the Finance Department. Not only will the SPL and training matrices go towards the company's Investors in People accreditation, but they will play a significant part in evaluating competency levels when the pay structure is reviewed later in 1999.

TPM Facilitator, Gordon Hill, planned and prepared each SPL with the TPM teams, with meticulous attention to detail, prior to the introduction of the full set of documents to the shopfloor. An excellent example of heeding the adage 'Failing to prepare means you are prepared to fail', Gordon Hill and his shopfloor colleagues have invested a great deal of time to ensure that Henkel gets it right first time, every time.

Quotable quote

> The OEE calculation is without doubt the single most impressive feature of TPM. For the first time it was possible to measure the total cost of non-conformance in a simple and straightforward way.
>
> Gordon Hill,
> **TPM Facilitator**

3.0 Benefits

Team focus

The two teams in the pilot study, conducted in the summer of 1997, realized valuable improvements after just four months.

The Fischbach team began with an OEE of 45 per cent, which they improved by 6 per cent to 51 per cent. An improvement target of an 80 per cent 'best of best' OEE has been set, which has a projected cost saving of £45 000 per annum.

Meanwhile, the CTA team raised their OEE from 37 per cent to 46 per cent. The best of best for this team shows that 74 per cent is possible. When realized, this will result in a reduced operating cost of £44 000 per annum.

When extrapolated across the site, realistic improvements in OEE have the potential to make a significant impact in reducing costs.

Although only in its early stages, the introduction of a TPM culture at Henkel and use of the OEE measure throughout the plant has already given the company some valuable benefits.

The introduction of the CMMS means that valuable data can be recorded, from machine speeds and product information to shift patterns and planned vs. actual stoppages. A detailed 'six losses' report is produced on a daily, weekly and year-to-date basis, highlighting every single loss, however small. Says Gordon: 'A loss of two minutes may seem insignificant in itself, but not when records show it is happening 500 times a week!'

The report is used to illustrate the most common losses both across the plant and by machine, giving focus and direction to TPM improvement teams and allowing detailed attention and action to take place. It also gives target OEE for specific improvements relating to individual losses, and thus projected forecasts of cost savings are generated. This is then approved by the Finance Department as a recognized cost-reduction programme.

Benefits

- Two TPM projects, launched in November 1998, are already projecting a manufacturing cost reduction of 20 per cent per annum.
- Every line in the factory now has an OEE benchmark.
- Every line now has improvement targets.
- Every project on each line now has its own set of improvement values with respect to availability, performance and quality rates.
- There is an awareness regarding the issues affecting the productivity of each line.
- The data emanating from the OEE module of the CMMS is helping to drive the company's continuous improvement projects.
- Henkel now has the ability to analyse production problems on a daily, weekly and year-to-date basis, allowing trends to be pinpointed and opportunities for improvement to be mastered.

4.0 The future for TPM

Although only in their first year of implementing TPM, the programme has already realized tangible financial results. Good and consistent planning means that the future will be based upon firm foundations and a determination to move forward to realize the full potential of a site-wide TPM and continuous improvement programme.

RHP Bearings

By Danny McGuire, General Manager, and
Kevin O'Sullivan, TPM Facilitator

1.0 Background issues

RHP Bearings in Blackburn, which manufactures cast iron bearing housings for a variety of uses from agricultural machinery to fairground rides, is one of seven RHP manufacturing sites in Europe owned by Japanese group NSK, the world's second largest bearings manufacturer. NSK acquired RHP in 1990, when the Blackburn site was under the imminent threat of closure because of high costs and the subsequent lack of competitiveness.

Employing 93 staff and producing 220 product types, RHP Bearings Blackburn has turned its fortunes around through the efforts of its workforce and the support of NSK. Total Productive Maintenance (TPM) was introduced to the site in 1993 and has since become the driver of all continuous improvement activities. As a result, RHP Bearings has reduced unit costs, increased productivity and attracted capital investment to the site – all under a no-redundancy agreement.

The site is now so flexible in terms of customer response that it can turn a product around in less than two days, compared to two weeks in the early 1990s.

The company's rise from the ashes towards becoming a world-class organization has been recognized by a series of prestigious awards, including Investors in People, the Business Environment Association, North West Quality Award and ISO 14001 environmental status (one of the first iron foundries in the world to achieve it). The site has also been used as a case study visit for last year's TPM5 Conference. All this has been achieved through the rigorous implementation of the TPM methodology.

2.0 Why TPM?

Although first introduced five years ago, TPM has only become an integral part of the company culture since it was reinvigorated eighteen months ago. Until then, previous efforts to drive TPM had failed because it was largely theoretical and the workforce failed to see its relevance to the everyday running of the plant.

Then in early 1996 WCS International was brought in to do a scoping study of the plant, run a four-day workshop in conjunction with Lynn Williams of the AEEU, and to launch and support two pilot TPM projects. This time, under the leadership of Plant Manager, Danny McGuire, and TPM Facilitator, Kevin O'Sullivan, TPM was made directly relevant to the jobs of the staff.

Operators were sent off to climb over their machines and log problems through a detailed condition appraisal, to establish a foundation for future TPM improvements.

TPM was piloted on two key machines, the PGM core making machine in the foundry and the Shiftnal sphering machine in the machine shop, using a detailed seven-step TPM implementation programme:

1 Collection and calculation of Overall Equipment Effectiveness (OEE) data
2 Assessing the six losses
3 Criticality assessment and condition appraisal
4 Risk assessment
5 Refurbishment plan
6 Asset care and best practice routines
7 Regular review for problem solving

TPM is currently implemented at the site by nine different TPM equipment teams, involving 60 per cent of the workforce.

TPM is applied to machines of all ages – from new to 30 years old – ensuring that older machinery is brought up to modern specification and newer machinery is kept in 'as-new' condition. TPM can also help the running of new machines in other ways. For example, the new shot blast machine runs at an OEE of just 60 per cent due to hold-ups in other parts of the foundry process. TPM has identified the external bottlenecks to allow the machine to work to its full potential.

The cross-functional teams include operators, maintainers, quality technicians and group leaders. These core teams are supported by Kevin O'Sullivan and can also draw on the skills of key contacts from other areas. For example, Quality Associate and Toolmaker, Alan Shaw, works for four teams as a key contact because of his specialist skills.

Each team has worked hard to develop a standard TPM routine for its respective machine, using the following methods:

- *Autonomous Maintenance System (AMS) boards*
  These mobile boards show a schematic of the machine which the operator then tags with labels to show losses affecting availability, performance and quality. The labels are then used to generate an agenda for TPM team meetings.
- *TPM step notices* Notices on each machine illustrate its current stage in the seven-step process.
- *Mainpac database* An in-house database is used to gather machine performance data and calculate OEE. The system is also used to assess the over-maintaining of machinery, where OEE results are consistently good but maintenance levels high.
- *Key performance indicators* Each team assesses itself according to progress and improvements in the following areas:

- Waste sand
- gas emissions
- *Kaizen*/continuous improvement
- Attendance
- Customer returns
- Lost time accidents
- Injurious accidents
- Audit conformance
- Product conformance
- Mainpac reports
- Activity boards

Each team has an activity board covering subjects such as milestone activities, for example schedule adherence (Just in Time), key performance indicators, training status, health and safety, today's quality actions and previous day's conformance report by production, scrap and target.

- *Total Manufacturing Concept (TMC),* A system which pinpoints actions for each machine, each month, under headings of Quality, Cost, Delivery and People.
- *Quarterly housekeeping audits.*
- *Shadow boards* Each area now has shadow boards for storing tools in an ordered and easily identifiable way.
- *TPM for Design* TPM for Design principles are applied when specifying new machines and processes.

To keep abreast of such a variety of routines, the teams have to dedicate substantial time to TPM activities, as well as flexibility to ensure that a meeting deferred is not a meeting abandoned. In spite of production pressures, TPM team members recognize that they have to 'take time to save time'.

3.0 Benefits

TPM has both a direct and an indirect effect on a production system. The direct effects are easier to assess and are directly quantifiable (see Table 11.1).

The indirect effects can be ascertained from more subjective measures, such as morale and absenteeism levels, which show themselves as an increase in the efficiency of the overall production system. Estimating cost savings involves looking at a variety of factors, from labour and materials costs to energy savings and increases in production capacity.

Indirect savings attributed to TPM include sharing of best practices, both across the shifts and with the site in general, and TPM teams solving problems that have a beneficial effect on unrelated areas, e.g. improving stock control systems.

The combination of indirect and direct effects at RHP Bearings Blackburn generated major savings. Major site-wide benefits from TPM activity were scored in the following areas:

- £400 000 running costs saved
- Unit cost reduced by 21 per cent
- Scrap reduced by 8 per cent
- Attracting increased capital investment currently at 15 per cent of turnover
- Customer returns reduced by 11 per cent
- Increased customer satisfaction
- Improved safety record
- Environmental and quality awards
- Improvement in staff morale

Table 11.1 Example of four TPM Teams

Direct TPM savings

Team	*Saving*	*Main benefits from:*
Beach Boys	£50 000	Set-up reduction
WRIST	£20 000	Set-up reduction
DREAM	£80 000	Cycle time reduction
Shiftnal	£60 000	Cycle time reduction
Total	£210 000	

Shiftnal sphering machine

The Shiftnal, a 26-year-old sphering machine used for the production of a range of larger castings, is one of the oldest machines at the plant.

The TPM team responsible for the Shiftnal has now developed a detailed refurbishment plan, having almost completed the TPM cycle. Through the implementation of TPM, the machine has been brought up to PUWER (Provision and Use of Work Equipment Regulations) standard, with the use of two button start, guarded off areas and asset care with a cleaning and preventive maintenance schedule.

Other key improvements include:

- Total cost saving OEE improved by 25 per cent. The team are now targeting areas of poor OEE by product and problem type.
- Cycle time improved by 6 per cent, equating to the production of an extra sixteen castings per day with a potential sales value of over £30 000. The improvement was implemented though the introduction of a sphere position switch, costing just a few pounds.
- Changeovers speeded up by the production of tool kits, so that all items required for each casting type are readily available and labelled.

A variety of other time-saving methods have been put in place. For example, to check the torque reading of the casting, the operator, Norman Jameson, had to lift the heavy metal casting into a vice set upon the workbench. At

Norman's suggestion, the vice was set down into the bench so that he can slide, instead of lift, the casting into it – an improvement costing £20 which has had a significant impact on health and safety by improving the ergonomics of the process.

PGM core machine

The PGM core manufacturing machine moulds resin-enriched sand which is then used in the casting process. Prior to TPM, there were chronic losses in the process, including unplanned downtime, quality and excessive waste.

A series of more than twenty improvements have been made by the TPM team to date, including:

- Making a bolt box, costing only £2, which has led to an annual time saving equivalent to £624
- Replacement of valves and gauges in the amine gas system
- Clearing out the sand pit that had formed at the base of the machine. Prior to attention from the TPM team, aptly named the Beach Boys, sand was allowed to build up to waist height before being cleared. As well as getting at the source of sand spillage, any surplus sand is swept away regularly and dropped tools no longer accumulate under the mound of sand
- Installation of a blow plate. An initial cost of £50 represents an annual cost saving of £936

Team achievements

- Total cost saving doubling the OEE
- Average set-up time reduction of 30 minutes per event
- 88 per cent reduction in scrap

The machine will be moved to a location which incorporates bulk storage of raw materials – a need highlighted by TPM activities. The project will also address new environmental legislation concerning emissions of the gas used as a raw material.

The new set-up will save hours in changeover times, due to the use of quick-release fittings and supplier involvement in changeovers, as well as dramatically reducing leaks.

Valued at £80 000, the project has received £30 000 funding from NSK – a testament to its commitment to the work of the Beach Boys team.

Quotable quotes

> The key to TPM is making it easy to do things right and difficult to do things wrong.
>
> **John Smith, Team Leader, Machine Shop**

> Ask the operator to get involved in maintenance and there will

be a pride of ownership that lasts way beyond completion of the actual maintenance task.

Bob Tormay, Group Leader, Foundry

Before TPM we were just issued jobs – no input asked for, no feedback given. Now we get feedback in our TPM meetings and have a much more responsible role because of the team-working.

Graham Wignall, Fitter, Beach Boys team

Immediate objectives

- £500 000 of further cost savings
- Reduce unit costs to give a total reduction of 35 per cent over two years to the end of 1998, enabling RHP to compete in Asian markets
- TPM applied to all key processes
- Reinforcing 5S/CAN DO discipline
- Trialling of new TPM database to run alongside Mainpac, which will record fault history. All twelve critical machines in machine shop to be recorded on database, from current three
- All site staff involved in TPM by the end of 1998
- Run awareness workshop with new team members
- Coaching existing teams
- Each team to have a mentor from management team
- Teams to have forum for sharing ideas
- Team responsibility for setting own objectives and sign-off required before moving to the next stage
- Standardizing of TPM activity boards

4.0 The future for TPM

RHP Blackburn's continued commitment to TPM is echoed by its mission statement for 1998: 'To secure our future through the development of our people, plant and processes, in order to achieve the company vision of becoming a world-class organization.'

A priority for this year is the retraining of team members. As TPM has developed at the plant, enthusiasm for results has meant that steps in the process have sometimes been skipped. The training will refocus the teams to the seven-step process and emphasize the need for obsessive attention to detail.

Just like the individual teams, the TPM development at RHP is evolutionary, reaching and then maintaining new levels of best working practices. As Plant Manager, Danny McGuire, concludes: 'The greatest asset we have at RHP Blackburn is our people. It is they who have made TPM work and it is TPM which has allowed their potential and enthusiasm to be tapped.'

Total Productive Mining
RJB Mining

By Grant Budge, Director of Mining Services, with help from Stuart Oliver, Chris Crouch and others named in the case study

1.0 Background issues

RJB Mining has made upwards of £500 million investment in its UK coal mining operations during the late 1990s following its acquisition of various mining interests, both surface and underground, from the old British National Coal Board.

However, whilst long-term, high-volume replacement contracts have been won from the UK's electricity generators, it has been at the expense of significant price containment.

Finite market volume and increasing environmental standards means a harsh reality where it is essential to gain real sustainable improvement in productivity and, hence, unit cost per tonne delivered.

2.0 Why TPM?

In addition to market-led pressures, we are always seeking ways to build on our excellent levels of teamworking, based on our beliefs that we will continue to:

- unlock our installed productive capacity
- by eliminating waste in all its forms
- by unlocking the positive energy of our people
- through involvement and sustained commitment
- to give ownership and pride

Hence the attractiveness of TPM, or Total Productive Mining as we prefer to call it.

3.0 The story so far

Seven sites are being initially targeted for a rolling programme of measures aimed at further improving the effectiveness of equipment and production processes at pits, surface mining sites and our central engineering workshops.

Following the successful launch of the Total Productive Mining scheme at Daw Mill Colliery, the focus has now moved to Harworth Colliery for commencement on a scheme which, in the few months since its inception, has identified numerous cost savings and improved working practices.

Six further sites will be identified during this year for a TPM programme which will embrace all significant RJB locations before the end of 2001.

Two 'facilitators' – Sean Kelley and Tim Marples – have been recruited to help roll out a programme which potentially could save the company millions of pounds, reduce production costs and improve safety. Sean, 28, previously worked for Elida Fabergé, part of the Unilever group, where he was a full-time TPM facilitator, while Tim, 37, was previously with Miller Mining, where he had valuable experience in problem-solving techniques and change management facilitation. Their responsibility within the organization will be to train, assist in implementation and provide continual practical support for TPM across the business, and to sustain an environment of continuous improvement.

Says RJB's Mining Services Director, Grant Budge:

> As was explained in the last edition of NewScene, TPM is about developing the business together and continuously focusing on the areas of lost potential within our processes.
> TPM is not designed to replace our current practices, but to support and develop them so that we can collectively improve the safety and performance of our workplace. It's not about working harder – it's all about working smarter.

While new to RJB, TPM is an internationally accepted business development strategy and has been used by a large number of companies, and a wide range of disciplines, over the past twenty years. It extends the 'team philosophy' across the shifts and disciplines, seeking to maximize efficiency and product quality through teamwork.

Being results-driven, TPM looks at the complete process, rather than individual elements, reviewing the entire cycle of events to identify and correct weaknesses and minimize losses.

4.0 It's all about teamwork

Time is money – and on Daw Mill's key 204s coalface, an additional 1 per cent improvement in equipment effectiveness could increase production by 2 per cent. It's here, 600 metres below ground in Shakespeare country, that Command Supervisor, Mark Emmett, and his boys are doing the business thanks to a forward-looking management style that's seen face teams grasp the nettle of change.

On Warwickshire Thick 204s, a 300-metre faceline taking a four-metre coal section with some of the highest rated equipment ever seen at a British mine, a core team charged with improving equipment effectiveness and planned maintenance systems has been formed – and from the support they are getting, they're producing results.

Total Productive Mining is no midsummer night's dream at Daw Mill. As Mark says:

> It isn't just a management thing. The fact that the process operates horizontally gives it the best chance of success.
> Men are being given a role they've never had before and there's no ignoring their commitment to the aim of TPM.
> TPM is a device to make things happen. We've got one or two sceptics out there who think it's a waste of time, but what better way to improve the way we do things than to have people at the working level looking at all of the issues and problems at their locations.
> After a close study of 204's stageloader we discovered, through analysis of production loss problems, that its running time was low. By monitoring delay times and causes, we pinpointed various problem areas with the Bingham net, blown fluid couplings, blocked chutes and chains being fast.

In the new 'Make it Happen' partnership, the core team of nine faceworkers, craftsmen and officials has thrashed out solutions, recognizing the valuable input of others to improve the efficiency of the equipment.

Adds Mark:

> As the doers they've vast experience built up from thousands of hours at the sharp end.
> Don't get me wrong, we looked at equipment delays in the past, but it was never in such depth, or with such a clear view.
> I've been very impressed about the work TPM can do and we are now preparing to take its message to the coal shearer.
> My manager, Keith Williams, has given me carte blanche to promote TPM. It means that I, and people like me, are better informed about issues like asset care which breeds good practice.

Adds Keith

> Management is moving from a directive role to that of a facilitator.
> It's part of TPM to empower people at different levels of the business to identify problems and act to solve them and so to strive for quality where they work.
> By working together in this change process we will achieve our aims to be a better supplier.
> When we succeed, we help our customers to succeed and through them the company succeeds – that's why we encourage teamwork.

5.0 Prep men in front line

Coal prep men have joined the front line in Daw Mill's new formula for creating change – a move that's dedicated to produce quality as well as quantity.

Wherever possible, the washery team is being involved in the day-to-day running of operations. It's no longer regarded just as the foreman's responsibility – the buck stops at every man.

According to the Marketing and Quality Control Manager, Peter Bottrill, Total Productive Mining is seen to be effective both in bringing changes and encouraging participation. He says:

> The men are becoming a vehicle for change through their commitment to the TPM concept. This isn't a passing fancy – we're asking if there's a better way of doing a task or using a particular piece of equipment.
> Total Productive Mining is an essential aid to efficiency. And everyone – from sampling to rapid loading bunkers – has a part to play.

Being invited to take responsibility in the running of the business gave birth to dramatic improvements on a crusher, about nine months old, where unwanted hessian bags with the 'run of mine' were taking their toll and hitting productivity.

Peter brought together the men associated with the crusher, which was clocking up delays equivalent to 10 per cent of the plant's standing downtime each week. 'Coal washing often had to stop because that part of the operation kept failing,' he said.

By pushing as much responsibility for improving the efficiency of the crusher down the line as he could, the 35-man team has grasped the opportunity with both hands.

With facilitator, Tim Rawlings, who has been a foreman for ten years, the team carried out time running monitoring as a fuller picture emerged of the delays being experienced at the plant, which is designed to process 600 tonnes of 'run of mine' coal an hour. The result is a refurbishment plan for the infeed to the crusher and installing bag catchers which have had an immediate and dramatic effect.

The team has regular meetings to review progress and is now preparing to take its quality message to other parts of the plant.

Says Peter: 'Because we put the problem across to the lads to solve themselves, they have developed a pride in their own plant and a sense of ownership.'

The knock-on result is a major clean-up programme. Real time is being given over to asset care, which has become the new buzz phrase on everyone's lips.

Just as important, management has provided the time and the means – and is letting the men get on with the job. The bottom line result is improvements in washing time being achieved.

As Tim says:

> Team work is being encouraged as never before. Our core team is a mixture of operators and maintainers who are working together and learning about each other's problems.

> After the core team comes to the end of its work, they know each other better and go on talking to each other and exchanging ideas.

Adds Peter:

> We've got to give the customer what he wants at the right price – and that means a reliable supply of consistently good quality coal.
>
> Gone are the days when a coal prep plant was there just to remove the dirt. We're no different to Cadbury or Toyota whose customers demand quality and we want Daw Mill's name to be synonymous with a good quality product so that when it comes to fighting in the market place, we make sure we win more than we lose.

6.0 The TPM approach cuts costly delays

Trend-spotting heading men are reducing costly delays by up to 10 per cent. The TPM approach has provided Daw Mill with vital information on machine reliability which is building an accurate picture of the day-to-day running of 301's tailgate.

A core team of five development men, a fitter and an electrician are dedicated to bringing improvements in the high priority drivage that's seen the barometer – performance effectiveness – rise. Put another way, 5.2 × 3.7 metre wide development using 29 bolts a metre for both roof and ribs and supported with steel has opened out 120 per cent of plan since TPM was introduced.

Says Command Supervisor, Mark Gee: 'While major breakdowns are dealt with immediately, the main aim is spotting trends which have the knock-on effect of a few minutes lost regularly here and there.'

After a close study of the drill rigs on the bolter miner, the trend-spotting team discovered, through analysis of production loss problems, that running time was low. By monitoring delay times and causes, various problems were pinpointed on the drill rigs, which were clocking delays the equivalent of one shift every week. The heading often had to stop because of failing hose seals and burst hoses.

Among the recurring hold-ups the team spotted was the delay with fitters getting spares for four different types of hose fittings. The problem was simply solved and valuable time saved by standardizing the hoses.

Many ideas on problem solving have come from the men, who are playing an important part in helping to record accurately what is going wrong.

The whole team is involved with decisions – more involvement means a greater interest in getting things done. They now have pride in the district and a sense of ownership.

For Mark Gee, the team's facilitator, who has been a supervisor for thirteen years, the new business environment at Daw Mill has taken some getting used to.

I've had to change the way I think about the business and that's what we're trying to encourage others to do.
I'm absolutely certain this is the way we have to go. Men can see their ideas are being heard and this has become a breeding ground for good habits in everyone's work.
Put simply, we are aiming at a change in developing attitudes which are more in line with commercial realities.
'The TPM philosophy is work smarter, not harder, by dealing with losses and delays on the machine and preventing reoccurrence by best practice development and effective asset care. It doesn't mean the driver will be working any quicker, but his productivity will increase through improved machine running time.
The bottom line is that man and machine become more efficient.
Old Will himself could't have put it better.

Index